Principles of welding tec'

In memory of Frank Copleston – friend, colleague, and fellow enthusiast for welding.

Principles of welding technology

Third edition

L. M. Gourd
Formerly Group Manager, Training Services, The Welding Institute

A member of the Hodder Headline Group
LONDON • SYDNEY • AUCKLAND

First published in Great Britain in 1980 by
Edward Arnold, a division of Hodder Headline PLC,
338 Euston Road, London NW1 3BH

Second edition 1986
Third edition 1995

© 1995 L.M. Gourd

British Library Cataloguing in Publication Data
A catalogue record for this book is available from the British Library

ISBN 0 340 61399 8

   3 4 5 6 7 8 9 10

Typeset in Times by GreenGate Publishing Services, Tonbridge, Kent
Printed and bound in Great Britain by J.W. Arrowsmith Ltd., Bristol

Contents

Preface

One hundred years after the first patent on arc welding was filed, the public image of welding is still that of a man behind a mask, creating bright lights, smoke, and sparks. For the TV viewer, this is the spectacular scene which is used to evoke the idea of heavy industry. For others, welding is associated with a garage mechanic repairing a damaged or rusted wing panel on a car. Even the trained engineer often has a restricted view of welding, limited to a large extent by everyday experience of one company's practices. Yet the term 'welding' covers a large number of techniques which can be used by the production engineer. Indeed, it is this very diversity which has made welding indispensable to manufacturing industry. So many of the things we rely on for modern living would not be practicable were it not for welding – from the construction of oil platforms and power plant, through volume production of cars, to the sealing of transistor capsules. It is therefore surprising, to say the least, that the technology of welding is so poorly understood by many engineers, technicians, and craftsmen. The effect of this is to be seen in poor-quality products and lost opportunities for significant improvements in productivity. In writing this book, I have tried to make some small contribution towards remedying this situation by providing an insight into the technology and by highlighting the various options open to the production engineer .

One of the problems of teaching welding technology at an introductory level is that it involves an unavoidable amount of description which can easily lose the student's interest. I have tried to overcome this problem by concentrating on the logic behind the existence of the various techniques. The book is intended to be read as a continuous text, and I hope that in doing so the reader will capture some of the intellectual challenge which comes from the involvement of many basic disciplines. This type of treatment has meant that the book does not follow any particular examination syllabus. It does cover, however, the theory content of current welding craft and technician courses (NVQ Levels 2 and 3) and would be useful for students on more general engineering courses at GNVQ and under-graduate level. The book will also be useful on industrial training and continuing education courses as the text deals with many topics included in the syllabuses devised by the European Welding Federation.

In the preface to the first edition I wrote that welding is a developing technology. The intervening years have confirmed this, and many techniques which were being developed then are now in common use. In particular, the development of power supply units for welding has already led to significant changes in the application of gas-shielded processes. Similarly, interest in the response of metals to welding has increased not only with the wider use of materials such as stainless steel and aluminium alloys, but also as a result of demands for better quality welds in conventional steel structures. The publication of a third edition has provided an opportunity to up-date various aspects of process technology and to provide a brief introduction to weldability. In revising the text, however, I have not lost sight of the fact that the majority of welding in the UK is still firmly based on manual skills, and I hope that this aspect has been presented in a way which will encourage welders to enhance their knowledge of the principles underlying their skill.

I would like to thank my many colleagues who have made helpful comments on the text and to acknowledge the help given by The Welding Institute with many of the illustrations. I would also like to thank the following organisations for supplying illustrations of items of their equipment: BOC Ltd (figs Bl(b) and B4), ESAB (UK) Ltd (figs 4.24, 10.6,10.21, and 10.24), and Thompson Welding Systems (fig 10.19). Material from British Standards BS 499: Part 1, BS 679 and BS 2573: Part 1 has been used by permission of the British Standards Institution, 2 Park St, London W1A 2BS, from whom copies of the complete standards may be obtained. Finally, I want to record my thanks to Russell Parry and his colleagues at Edward Arnold (Publishers) Ltd for their patient and invaluable help in preparing the text for printing.

L.M. Gourd

1 Which method of joining?

1.1 The development of joining techniques

Though metals have been used by man for many thousands of years, no one is certain how the first useful metal was produced. It may be that metal nuggets left on the surface of the earth by meteorites were found to have useful properties. More likely, an early inhabitant of an area containing copper-bearing minerals inadvertently heated these ores in a open fire. Under suitable conditions this would have produced a lump of impure copper which could be hammered into shape. Whatever their origins, the early use of metals has been confirmed by the discovery of implements made from a copper alloy (bronze). Axes, spearheads, and ornaments have been excavated from the sites of primitive settlements, and archaeologists have been able to show that they were made and used in a period which we now call the Bronze Age.

Bronze is an alloy of copper and tin, and one of its characteristics is that it can be shaped to give a good cutting edge. This would have made it very attractive to the late Stone Age hunter, who had been forced to do the best he could with flint axes and spears. Also, the introduction of bronze into domestic tools meant that constructional techniques, especially with wood, were improved.

The uses to which the newly found metal could be put were limited by the fact that the technology available could not offer techniques for producing large items entirely in bronze. This was not a handicap in the case of an axe or a spear, since there were operating reasons for using a resilient material such as wood for the shaft. Various methods were devised for joining the head to the shaft which satisfied the requirements of the time, but the problem of achieving acceptable metal-to-metal joints was to remain unsolved for many centuries. Indeed, apart from the development of forge welding, which is normally attributed to the Syrians, about 1400 BC, the inability to join small pieces of metal to produce a larger or more complex component hampered engineering progress even up to the beginning of the nineteenth century. Admittedly, casting techniques were exploited to the full, as in the production of bells and cannons, but the metal tended to be brittle and lacked the strength of that forged under the blacksmith's hammer. But, forged components tended to be small and needed to be joined together. The developments that did take place were largely in response to military requirements. A good example is the riveting of suits of armour, which enabled damaged panels to be replaced effectively. It needed the industrial revolution, with its dependence on machinery, to provide the incentive which led to the introduction on a commercial scale of bolting, riveting, soldering, and finally welding.

Today there is a large number of joining techniques available, and the problem is not how to join but how to select the best method of joining. Whereas bronze-age man simply had to chose between thongs and wedges, a design engineer today may well find that four or five techniques appear to be equally suitable. Each method has its own attributes, and a number of aspects must be evaluated if the final choice is to be sensible. The relative importance of such factors as strength, ease of manufacture, cost, permanency, corrosion-resistance, and appearance depends very much on the specific application.

The designer of a bridge looks for a means of producing joints in plates which will carry the varying loads imposed by vehicles travelling across the deck. While it would obviously be desirable to carry out the joining operations as quickly as possible, this might not be a prime consideration in all cases – the fact that some of the joints will have to be made on site could well be an overriding factor. Thus, even though welding is widely used in fabricating girders, critical site joints connected with these girders may be bolted in the interests of better control of quality and ease of working in difficult locations (fig 1.1).

Bolted connection made on site

Fillet welds made in fabricating shop

Fig. 1.1 *Fillet-welded girder with bolted web connection*

By contrast, the manufacturer of cars travelling across the bridge will be looking for joining techniques which can be incorporated into production lines, combining speed with reproducibility and reliability. As far as possible they must be capable of being used by semi-skilled operators, since a large labour force is usually required. This, in turn, demands that the production sequence is established and controlled or monitored by specialist personnel.

1.2 Manufacturing methods

Any discussion of joining presupposes that the decision to assemble the unit from a number of parts is correct. This may not always be a valid assumption since there are many viable alternatives. Casting, forging, pressing, bending, extrusion, spinning, and machining can all be used to produce both large and small components in a variety of metals.

These processes are by no means the only ones available, but they illustrate the range of manufacturing techniques on offer to the modern engineer. It is important to recognise that they are not mutually exclusive, and an economic production sequence will normally incorporate two or more of these methods linked to the judicious use of suitable joining techniques. We must also be careful to see joining – especially welding – not as something existing in its own right but as part of a corporate activity that takes pieces of metal and converts them into the product which the customer wants. It is also wrong to think only of sheets and plates when considering joining – frequently we are faced with the need to produce connections to or between castings, forgings, or extruded sections (fig. 1.2). This must be borne in mind as we review the various joining systems and subsequently discuss welding in detail.

➡ Indicates direction of flow

Fig. 1.2 *Cooling panel made by welding together extruded aluminium panels*

1.3 Mechanically formed joints

Simple butt and corner joints can be made by folding and interlocking the edges of sheet-metal components (fig. 1.3). A large amount of deformation is needed to produce a secure joint, and this method is applicable only to material which has good ductility and can be bent around small radii. The strength of the joints is acceptable for many applications such as the manufacture of sheet-metal containers and ducting. Where higher strength or better leak-tightness is required, the folds can be filled with solder, as in food cans.

3

Fig. 1.3 *Mechanical lock joints*

1.4 Lap joints

Lap joints are used in most units fabricated from sheet metal (fig. 1.4); they are less frequently used to join plates. There are three basic types – butt, T, and corner – each of which can be joined in two different ways: we can either connect the sheets only at points along the length of the joint or provide a continuous bond over the complete interface.

Various methods are used to produce a local connection between the two parts. In structural work, bolted lap joints are common since they are easy to assemble

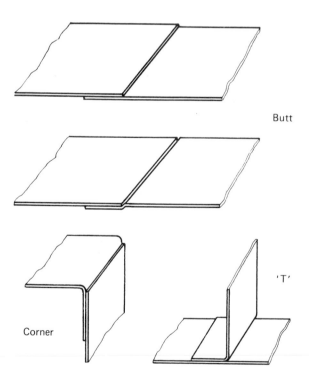

Butt

'T'

Corner

Fig. 1.4 *Types of lap joint*

on site or in the shop, provided the holes have been accurately drilled. Also, if high-strength friction-grip bolts are used, appreciable loads can be transmitted.

Another common technique used to secure lap joints is riveting, which was at one time the mainstay of the shipbuilding industry (it has now been replaced by welding). Rivets are widely used in sheet work as they readily give a joint of good strength and acceptable appearance. The skin panels on an aircraft contain many examples of the skilful use of rivets to provide joints which can withstand high stress. Blind or pop rivets are of particular interest where access is limited to one side of the joint.If the thickness of the material is less than about 2 mm, local connections can be made at the interface by fusing slugs of metal (fig. 1.5). The most common example of this technique is resistance spot welding, which is described in detail in Chapter 11.

Fig. 1.5 *Spot-welded lap joint*

Both riveting and spot welding produce permanent connections, whereas bolting enables the joint to be dismantled for repair or modification. Bolting tends to be slow for sheet-metal work and is frequently replaced by the use of self-tapping screws or threaded fasteners.

None of the techniques so far discussed gives a leak-tight joint without the use of a sealant. This problem is eliminated if we can bond the sheets together across the complete area of the overlap; at the same time, the amount of overlap can be reduced for the same strength. There are three possible ways of achieving this: soldering, brazing, and adhesive bonding.

Soldering

In soldering, a small gap between the sheets at the overlap is filled with a low-melting-point lead-tin alloy known as a solder. If the surfaces have been cleaned with a flux, the solder wets the metal and produces an intermetallic bond. If the gap is controlled to about 0.08 mm, the solder is drawn into the joint by capillary action, to give uniform filling. The melting point of the solder ranges from 183 to 275°C, depending on composition, and the joint is heated to the appropriate temperature by a soldering iron or gas flame (fig. 1.6).

Fig. 1.6 *Lap joint for soldering, brazing, or adhesive bonding*

Brazing

Brazing is similar to soldering but uses fillers having higher melting points (450 to 800°C). Brazed joints usually have better strength than soldered ones, but the need to heat to higher temperatures can pose problems, especially as oxidation and discoloration of the metal surfaces can occur. Various heating methods are used, such as gas torch, furnace, molten-flux bath, induction heating, and resistance heating.

Adhesive bonding

Compared with soldering and brazing, this is a relatively new technique, but it has found rapid acceptance due to its ease of use. As with soldering and brazing, a small gap is used. In this case it is filled with adhesive which forms a surface bond with the metal. The type of adhesive is varied to suit the particular requirements of the joint.

A lap joint loaded in tension rarely experiences simple shear (fig. 1.7) – the applied forces are not in line but are displaced, with the result that the joint tends to rotate as the forces try to achieve alignment. The use of butt straps enables the sheets or plates to be placed edge to edge so that the applied forces are acting in one plane. Such joints are more expensive to produce, require more material, and are heavier than the simple lap. In the majority of cases, a welded joint offers a better solution by providing a direct connection between the abutting edges of the parent metals.

(a) Single-lap joint subjected to tensile load

(b) Use of butt straps

Fig. 1.7 *Lap joints under load*

1.5 Welded joints

Most welded joints are made by melting the metal on each side of the joint line. The molten metal combines to form a weld pool, which on solidification becomes a metallic bridge between the two components of the joint. The sequence of events can be seen in fig.1.8a which shows the cross-section of a joint in sheet metal. In the case of a joint between two pieces of metal which are arranged at a right angle (fig. 1.8b), the metal melted from the surface of the components would not be sufficient to form a bridge. Additional metal, known as filler, must be added to provide a metallic bridge which has adequate strength.

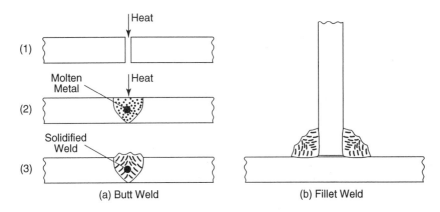

(a) Butt Weld   (b) Fillet Weld

Fig. 1.8 *Formation of the fusion welded joint*

Unlike the other methods of joining discussed so far, the success of fusion welding depends very much on the skill of the person operating the process. Bolting, riveting, soldering, brazing, and indeed spot welding involve operations which can be standardised and adapted to machine production and control. While this can sometimes be done with fusion welding, a number of operating factors rule out the widespread use of automatic welding. This is not necessarily a disadvantage, however, since the craft skill possessed by the welder provides a number of versatile techniques which can be used to fabricate products ranging from smallbore thin-walled tube systems to ships and bridges. Of equal impor-tance is the ability to accommodate one-off fabrications and to vary the type of work without having to invest in a major re-equipment programme.

Although it is true to say that many engineering advances such as the jet engine and the generation of nuclear power would not have been possible without welding, it is only one of a number of joining methods. Various advantages can be claimed for its use, but it must be judged on its merits in comparison with the other techniques we have discussed. Welding should be used only when it is the cheapest method of fabrication or it offers special technical advantages.

As a technology, welding has progressed rapidly in the last three decades, and we need to understand the principles involved if it is to be used effectively. In this book, we will be exploring the background to welding, the way in which the processes work, and the factors that govern the production of an acceptable weld.

2 Methods of welding

2.1 Types of welded joint

It is often extremely difficult to give accurate but brief definitions of common everyday things. This is very true of welding – daily we use equipment which rely

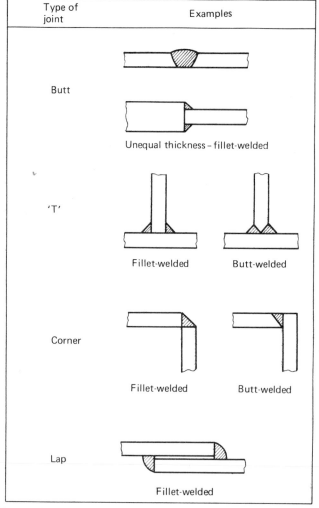

Fig. 2.1 *Types of welded joints*

on welded joints for their satisfactory operation, but rarely do we stop to consider the essential characteristics of the welds themselves. If we did, it would immediately become apparent that there are very many types of weld, which seem to vary according to the type of joint used. However, closer investigation would reveal that both the welds and the joints can be categorised into groups. We would discover that there are four basic types of joint: butt, T, corner, and lap (fig. 2.1).

The butt joint is characterised by the fact that the edges of the components are abutted and the load is transmitted along the common axis. This joint is of particular importance in fabrication, being used, for example, to join lengths of pipe, plates in ships' hulls, and flanges on bridge girders (fig. 2.2).

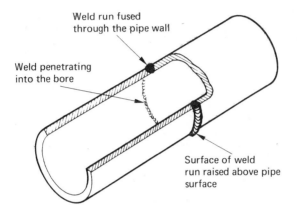

Weld run fused through the pipe wall

Weld penetrating into the bore

Surface of weld run raised above pipe surface

Fig. 2.2 *Butt weld between two lengths of pipe (sectioned to show bore)*

The T joint is probably the most commonly used connection in fabrication. Typical examples are at the flange-to-web junction in a plate girder, stiffeners welded to a panel, branches attached to a main pipe, and lifting lugs (fig. 2.3). The joint can be made either with no penetration along the joint line (using 'fillets' of weld metal to provide the load-carrying connection) or with bonding across the interface, i.e. butt-welded.

The corner joint can similarly be butt- or fillet-welded according to service requirements. Corner joints are normally associated with box sections, but it is worth noting that, from the point of view of both design and production, flange-to-pipe connections represent very important examples of corner joints.

Finally, the lap offers an interesting variant because it can be used in butt, T, and corner joints. Most commonly, the bonding in a lap joint is over only a small area of the interface, either as a number of spots (fig. 2.4) or as a narrow strip along the length of the joint. Lap joints are mainly used in sheet fabrication, and common items such as cars, washing machines, refrigerators, etc. contain many examples.

An important feature of all the joints discussed above is the need for a bond which will transmit loads. This means that between the two components there must be a continuous metallic connection – i.e. a metallic bridge – which will be strong enough to withstand the stresses applied in service.

9

(a) Beam-to-column
connection

(b) Branch pipe
welded to main pipe

(c) Stiffeners welded
to a panel

Fig. 2.3 *Examples of T joints which have been fillet-welded*

(a) Spot-welded

Bonded interface

(b) Fillet-welded

Fig. 2.4 *Examples of lap joints*

2.2 Weld formation

Consider the definition of a weld, and in particular look at the definition in British Standard 499: part 1 'Welding, brazing and thermal cutting glossary':

> A weld is a union between pieces of metal at faces rendered plastic or liquid by heat or by pressure or by both.

The first part of this definition follows from our discussion above: the union is the continuous load-bearing metallic bridge to which we have referred. But what is the significance of the reference to rendering the faces plastic (i.e. permanently deformed) or liquid? To understand this, consider how the union is achieved.

Sufficient strength can be produced in a welded joint only by interatomic bonding, and the prime function of the welding operation is, therefore, to provide links between atoms at the interface of the joint. For these links to be formed, two conditions must be satisfied. First, the surfaces must be in intimate contact: they should be atomically flat so that when they are brought together the gap between the respective surface atoms will be about equal to the atomic spacing within the

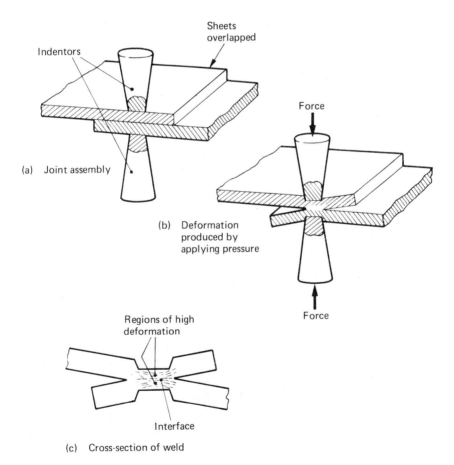

(a) Joint assembly

(b) Deformation produced by applying pressure

(c) Cross-section of weld

***Fig. 2.5** Cold pressure spot welding*

metal, i.e. 1.24×10^{-10} m. Even polishing to a high finish with diamond dust is unlikely to produce this degree of flatness. Second, the surfaces must be metallurgically clean – any molecules of grease, paint, moisture, oxygen, or nitrogen present on the surface will prevent the metal atoms from uniting, even if intimate contact is made. If a practical welding system is to be based on the idea of simple interatomic bonding, a means of bringing the surfaces into intimate contact and, simultaneously, dispersing the surface contaminants is essential.

2.3 Cold pressure welding

A way of achieving intimate contact and dispersal of contaminants is to force the surfaces together (fig. 2.5). Under pressure the surfaces deform, breaking up the contaminants and bringing areas of clean metal into intimate contact. The pressure required to disperse the contaminants leads to appreciable reductions in the thickness of the workpieces. Optimum joint strength is obtained at the threshold deformation level, whose value depends on the metals being welded (Table 2.1): the softer the metal, the lower the deformation required to initiate welding at room temperature. The deformation hardens the metal in the weld area.

Cold pressure welding is used to a limited extent to make welds between aluminium cable and connectors and for specialised applications such as welding caps to tubes (fig. 2.6), but it is usually difficult to accommodate the amount of deformation required for the welding of commercial alloys.

Fig. 2.6 *Fitting caps to tubes by cold pressure welding*

12

Table 2.1 Threshold deformations for welding

Metal	Threshold deformation (% reduction in thickness)
Lead	10
Tin	15
Aluminium	40
Copper	45
Iron	65

2.4 Hot pressure welding

Greater success can be achieved if the metals are heated during the welding operation. Raising the temperature reduces the value of the threshold deformation, and a number of successful hot-pressure-welding techniques have developed. Probably the oldest of these is forge welding, which has been in use by blacksmiths since about the year 1400 BC. In this process, the wrought iron or steel bars which are to be joined are heated to about 1350°C. At this temperature the iron oxides on the surface are melted and, when the components are hammered together, the molten oxides are squeezed out of the joint. Bonding then occurs at relatively low deformation levels.

Forge welding is now mainly used for craft work, but the principle of heating the component to make it easier to achieve a weld is also used in a number of modern developments of hot pressure welding. The method of heating is not critical, since the prime object is to raise the interfaces to a temperature at which the threshold value falls to about 25%. Since the hot metal is more plastic than it would be in cold pressure welding, the force which is required is smaller and it is possible to produce welds in hard metals such as steel. Unlike with a cold pressure weld, the material in the weld area is not hardened.

Of the various methods used to heat a joint for hot pressure welding, three of the most successful are gas heating, resistance heating, and induction heating.

Fig. 2.7 *Oxy-acetylene pressure welding of small-bore pipes*

Gas heating

A ring burner is arranged around the joint, and mixed oxygen and acetylene gas is supplied to burners spaced uniformly around the circumference (fig. 2.7).

Resistance heating

This relies on the heating effect of a current flowing through a resistance (fig. 2.8). For pressure welding, the components, which are usually round bars, are held in clamps and a high current is passed along the work through the interface. Although the abutting ends of the bars are held in contact under pressure, the oxides at the interface offer a resistance to the passage of the welding current and heat is generated. The rate of heating depends on both the current and the resistance, and the latter is to some extent influenced by the magnitude of the end pressure which brings the faces into close contact, thus increasing the area which can conduct the electricity.

Induction heating

This involves the passage of a high-frequency current through a coil surrounding the components to be welded (fig. 2.9). The magnetic field produced in the coil induces eddy currents in the steel work pieces, causing them to heat up.

Hot pressure welding is by no means a universally applicable technique, especially as it is usually restricted to bars, rods, pipes, and narrow strips. It is also difficult to use for 'one-off' work, since a number of trials must be made to ascertain the optimum combination of temperature and pressure.

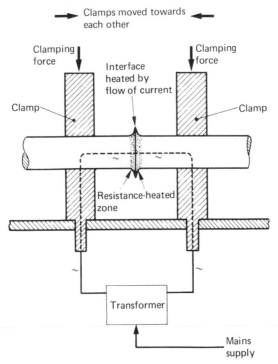

Fig. 2.8 *Principles of resistance butt welding*

motor, the other fixed. The two parts to be joined are clamped in the chucks and one part is rotated (fig. 2.11). This rotated component must be round in cross-section, but the part held in the fixed chuck can be a matching section or flat.

When the rotating chuck reaches the welding speed, the parts are brought into contact under a light axial load. As the abutting faces rub together, friction between them generates heat and localised hot plastic zones are produced. With the end load maintained, heat continues to be generated until the whole interface has reached a uniform temperature. At the same time, the plastic metal starts to flow outwards towards the periphery, carrying with it any oxides present at the joint face. When sufficient heating has occurred, the relative rotation of the parts is stopped rapidly and the end load may be increased. The result is a forged pressure butt weld having an excess-metal flash which can be removed by machining (fig. 2.12). Weld times are short, being around 20 to 100 seconds.

The operating parameters for friction welding must be determined by trial and established before production welding begins, so it is clear that the process is not really suited to one-off fabrications. Where there is a repetitive element, however, it offers considerable potential and its attraction to the producer of

Rotating chuck

Thrust cylinder

Sliding chuck

Brake

Motor

Rotation

Rotation — Force

Rotation — Force

Flash — Force

friction-welding sequence

16

Force

Coil
carrying high-
frequency current

Joint area he
by induced
currents

Force

Fig. 2.9 Using a high-frequency current *to heat the interfac*

2.5 Friction welding

Without doubt, the most successful development in the f
has been that of friction welding. The machine use
somewhat like a large lathe (fig. 2.10) fitted with two

Fig. 2.10 Typical friction-welding machi

Fig. 2.11 F

Fig. 2.12 *Cross-section of a friction weld*

engineering components is increased by the fact that friction welding units can be readily incorporated into workshops alongside lathes, milling machines, and similar machine tools. An example of friction welding in a high-volume production line is the manufacture of axle casings for cars and heavy vehicles (fig. 2.13).

Fig. 2.13 *Friction welding hub-end forgings to the axle casing for a heavy vehicle*

17

The basic friction welding process, as described above, suffers from a major constraint. One of the components of the joint must be rotated to provide the energy needed to generate heat. This can often pose operational problems as, for example, when welding long lengths of pipe. A variety of solutions have been devised to solve this type of problem. One approach is illustrated in fig. 2.14(a) and (b). A rotating wedge-shaped insert is placed between the abutting edges of the joint which are being pressed towards each other. In this technique, known as *radial friction welding*, heat is generated at two faces and plastic deformation occurs as it would in the conventional process. In another variant, one of the components is made to move to and fro by means of a vibrator attached to it. The fact that the movement is linear at the interface does not alter the mechanics of heat generation by friction. This technique is limited by the power available from the vibrator and, thus, is confined to the joining of small components.

Fig. 2.14 (a) *Joint setup, motion, and force for radial friction welding*

Fig. 2.14 (b) *Radial friction welding of pipes: completed weld*

18

Fig. 2.14 (c) Surfacing by friction welding

A quite different application of friction welding is to be found in surfacing (fig. 2.14(c)). The part to be coated is moved under a rotating rod of surfacing material which is pressed against the surface. The metal is bonded onto the surface in a series of overlapping pads.

2.6 Principles of fusion welding

All the pressure-welding systems described in sections 2.3 to 2.5 have limitations. In cold pressure welding, large amounts of deformation are required to achieve satisfactory bonding, while in friction welding the machinery for joining any except the smallest components is substantial and expensive. Probably the most critical limitation is that posed by shape and size. In fabricated structures, the joints which need to be welded more often than not consist of relatively long plate or sheet edges. These are not suitable for pressure welding, mainly because very large forging forces would be required. This would create problems in designing clamps to transmit the force and in accommodating the deformation which would take place. An alternative method of welding must be used for joints of this type. At the end of the last century, engineers, who were trying to devise new production techniques to maintain the impetus of industrial development, discovered that good welds could be achieved by the use of controlled fusion.

We can understand the principles of fusion welding by looking at the melting of a small area on the surface of a metal plate (fig. 2.15). As the solid metal reaches its melting point, oxide films which existed on the surface of the plate are disrupted and are either dissolved in the molten metal, as with iron or copper, or float on the surface, as with aluminium for example. This means that we have achieved the metallurgically clean surface mentioned on page 11.

There are now two regions at the joint line. The first is the solid metal which, although hot, still retains the ordered lattice structure which is a characteristic of metals. Examination of a polished and etched metal surface through the lens of a microscope shows a structure made up of a network of lines. These are the boundaries between areas or grains within which the atoms are arranged in

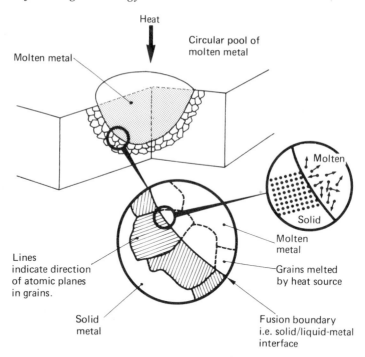

Heat

Circular pool of
molten metal

Molten metal

Molten

Solid

Molten
metal

Lines
indicate direction
of atomic planes
in grains.

Grains melted
by heat source

Solid
metal

Fusion boundary
i.e. solid/liquid-metal
interface

Fig. 2.15 *Formation of molten pool*

regular lines – the lattice structure. The boundary represents a change of direction
of the lines of atoms. The drawing in fig. 2.15 shows only a two dimensional
model, but in reality the atoms are arranged uniformly in all three dimensions.
The second region is the molten pool in which the lattice structure no longer
exists and the atoms are forming random, mobile patterns. If we could examine
the interface between the solid and liquid metals, it would seem as if the lattice
structure within each grain had been cut, leaving the end of each row of solid
atoms exposed to molten metal. We would also observe that there are no oxides
or other contaminants present – in other words, the solid metal at the interface
would be metallurgically clean. If the molten pool is allowed to cool, solidification takes place and appears to progress in layers from the solid/liquid-metal
interface towards the centre (fig. 2.16). Closer inspection would show that the
mechanism of solidification is far more complex. As the temperature of the pool
starts to fall, the molten atoms lose their energy. Since the heat is being conducted
away from the pool by the mass of the plate, i.e. the heat is flowing through the
solid/liquid interface, the atoms in this region are the first to look for permanent
sites. The solid atoms at the interface have spare bonds available to which the
solidifying atoms can attach themselves, so extending the lattice plane into the
molten pool. Thus a protrusion is produced to which more atoms become joined,
in this way extending the lines of solid metal into the weld pool.

Before long, atoms start to attach themselves to the sides of these protrusions,
with the result that lateral growth of the solid-metal 'finger' takes place and
continues until neighbouring solid regions meet at a boundary. The fingers of
solid metal are known as dendrites. Once this dendritic pattern has been established,

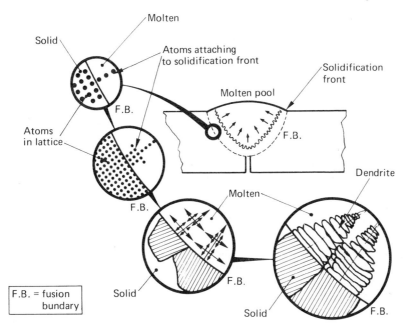

Fig. 2.16 Stages in solidification and formation of dendrites

the solid/liquid interface assumes a jagged or sawtooth profile and advances progressively towards the centre of the pool, until all the liquid metal has solidified. The volume which had originally been melted now consists of a collection of dendritic grains bonded to each other. Where neighbouring dendrites meet, the lattice planes are at an angle to each other and the junction forms a continuation of the grain boundary from the solid metal.

The localised bond made by melting just one spot is of limited application, and for most fabrications we need bonding along the complete length of the joint line. This can quite readily be achieved by moving the heat source along the joint line. In this way a series of overlapping pools is in effect produced, and progressive melting and solidification occurs at the leading and trailing edges of the weld pool respectively. Such a weld pool is no longer completely circular but has a round leading edge and an elliptical trailing edge (fig. 2.17). Solidification occurs at the rear of the pool only when the heat source has moved forward so that the flow of heat into the plate reduces the temperature of the trailing edge to the solidification point of the metal. The mechanics of grain growth are the same as those described for the stationary circular pool, except that the dendrites point upwards and forwards in the direction of the centre line (fig. 2.18).

2.7 Heat sources for fusion welding

The prime requirement for success in fusion welding is to have a source of heat which is capable of creating localised fusion. The fusion must be restricted to the ' joint line and should not be more than that required to achieve bonding. If the weld pool becomes too large, it is difficult to control and an unsatisfactory weld is produced. There are also other reasons why the overall heating should be

Solidified weld
with ripples showing
shape of weld pool

Leading edge
being
melted

Solidification
at trailing
edge

Molten
pool

Fig. 2.17 *Progressive melting and solidification in a welded joint*

restricted to a narrow band along each side of the weld. For example, metallurgical properties of the joint (Chapter 7) and distortion (Chapter 10) are related to the cycle of heating and cooling in the area of the joint.

A heat source must have a number of basic attributes if it is to be usable in practical welding situations:

a) The area heated directly by the heat source must be small so that the weld pool can be limited to a manageable size. Sources which heat a large area, such as those used for brazing and soldering, are not suitable for fusion welding.
b) The heat source must operate at a temperature significantly higher than the melting point of the metal being welded. If the differential is small, the rate of heat transfer is low and heat is spread over a wider area of the joint before fusion is achieved.
c) There must be an adequate heating capacity. The total amount of heat, or the rate at which heat is required, depends not only on the physical properties of the metal but also on the joint configuration and dimensions.

Fig. 2.18 *Etched section of weld, showing grains pointing in the direction of travel*

Fig. 2.19 *Oxygen-fuel-gas flame*

d) The heat source itself must be capable of regulation, so conditions can be set to suit the joint. These conditions must remain constant throughout.

The most common heat sources used in current industrial welding practice are:

1) an oxygen – fuel-gas flame,
2) an electric arc,
3) resistance heating at an interface,
4) and resistance heating in a slag bath.

An oxygen-fuel-gas flame

By burning a mixture of oxygen and a fuel gas at the outlet of an orifice in a tube or nozzle, it is possible to achieve quite high temperatures (fig. 2.19). Unfortunately, with most oxygen-fuel-gas mixtures these temperatures are too low for the welding of anything other than a few low-melting-point metals such as lead, zinc, and tin. The one exception is provided by acetylene. When mixed with oxygen in the correct proportions, this gas burns with a flame temperature of about 3100°C, which is adequate for many welding applications.

An electric arc

Normally the flow of electricity is associated with solid conductors such as wires. An electric current can flow across a gap between, say, a metallic or carbon rod and the surface of a plate (fig 2.20). In doing so it gives off a bright light but, more importantly from the point of view of welding, it can generate an appreciable

Fig. 2.20 *Electric arc*

Fig. 2.21 *Resistance heating at an interface*

amount of heat which can be used to melt metal in a small area such as a metallic joint. A low-voltage (20 to 40 V), high-current (30 to 1000 A) electric arc between the end of a rod electrode and the flat surface of a plate constitutes a very effective heat source which forms the basis of the majority of the techniques used to weld plates, pipes, and sections.

Resistance heating at an interface

When discussing pressure welding, we saw that a butt joint between two pieces of rod can be heated, to make the abutting ends plastic, by passing a high current through the interface. In that operation there was no intention of reaching the melting point of the rod material, but, if the current had been allowed to flow for a longer time or if the current density had been increased, the faces might well have melted. This principle can be applied to the production of a fused spot at the interface of two overlapping sheets (fig. 2.21). If, after melting has occurred, the spot is allowed to solidify, there will be a continuous metallic bridge over a limited area of the interface, i.e. a spot weld.

Fig. 2.22 *Resistance heating in a slag bath*

Resistance heating in a slag bath

An alternative way of using resistance heating is to pass a high current through a bath of molten slag (fig. 2.22). Mineral or ceramic slags are usually good insulators when cold, but they conduct electricity when they are melted. Their electrical conductivity is low, which means that they offer a resistance to the flow of current, and there is therefore considerable heat generated in the body of the molten slag. This heat can be utilised to melt metal plates in contact with the slag bath. Although widely used in metallurgical extraction processes, slag-bath heating features in only one major welding process: electro-slag welding and, its variant, consumable-guide welding.

2.8 Protection of the weld pool

All the heat sources discussed above produce a fused region having a temperature well in excess of the melting point of the metal being welded. If the weld pool, once established, is exposed to open atmosphere, it absorbs oxygen and/or nitrogen and/or hydrogen, depending on the composition of the parent metal. Small quantities of gas either dissolved in or combined with weld metal can usually be tolerated without detriment to the properties of the joint. In general, however, the presence of significant amounts of dissolved or combined gases is undesirable, since it often results in the formation of gas pores or voids in the completed weld. In addition, the properties or integrity of the joint can be impaired. For example, the ductility of aluminium weld metal is affected by the existence of oxide films at the grain boundaries; the electrical conductivity of copper is reduced by an increase in the oxygen content; dissolved hydrogen in high-strength-steel weld metal can lead to cracks in the joint (page 109).

Any viable welding system must include some method of preventing atmospheric contamination. In current practice two basic techniques are employed. From the point of view of historical development, the first of these is to blanket the weld pool with molten flux, forming a slag layer which is impervious to the passage of gases. The simpler alternative is to replace the air in the vicinity of the heat source and weld pool by a gas which does not react with the molten metal and is therefore in this sense inert.

2.9 Typical fusion-welding processes

It can now be seen that for each fusion-welding process we must consider two aspects: firstly the provision of a suitable source of heat and secondly the need to protect the weld pool against atmospheric contamination. In the following pages, profiles of some of the commonly used fusion-welding processes are given. The collection is not intended to be exhaustive; the aim is only to illustrate those systems which have made a major contribution to the success of modern fabrication engineering. Other, less widely used processes could be analysed and presented in the same way.

Oxy-acetylene welding (figs 2.23 to 2.25)

Alternative name Gas welding

Type of operation Manual

Heat source Fuel-gas flame

Shielding Products of combustion; flux for metals other than steel

Heat input See Chapter 3

Mode of operation An oxygen and acetylene mixture is burnt at the tip of a specially designed nozzle which is fitted to a torch body. The welder uses the flame to melt the parent metal to form a weld pool. Filler metal, if required, is added separately by manual feeding of a wire into the leading edge of the weld pool. The welder moves the torch to achieve uniform progressive fusion.

Typical applications Light fabrications such as ventilation ducts; small-bore pipework for heating and chemical systems; motor-vehicle repair.

Fig. 2.23 *General view of oxy-acetylene welding*

Oxygen/
acetylene
mixture

Products of
combustion protect
weld pool.

Filler-
metal rod

Torch
nozzle

Completed
weld

Weld pool

Typical torch

Fig. 2.24 *Schematic diagram of oxy-acetylene welding*

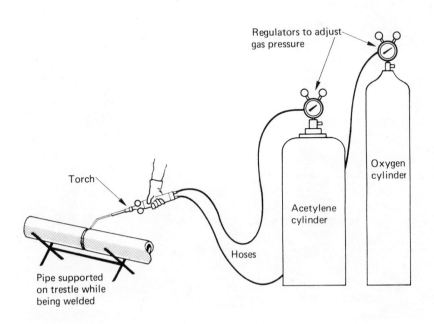

Regulators to adjust
gas pressure

Oxygen
cylinder

Torch

Acetylene
cylinder

Hoses

Pipe supported
on trestle while
being welded

Fig. 2.25 *Equipment for oxy-acetylene welding*

27

Tungsten inert-gas (TIG) welding (figs 2.26 to 2.28)
Alternative names Tungsten arc gas-shielded (TAGS) welding; gas tungsten arc welding (GTAW); Argonarc welding (BOC Ltd trade name)
Type of operation Manual
Heat source Arc
Shielding Inert gas
Current range 10 to 300 A
Heat input 0.2 to 8 kJ/s
Mode of operation An arc is established between the end of a tungsten electrode and the parent metal at the joint line. The electrode is not melted and the welder keeps the arc gap constant. The current is controlled by the power-supply unit. Filler metal, usually available in 1 m lengths of wire, is added to the leading edge of the pool as required. The molten pool is shielded by an inert gas which replaces the air in the arc area. Argon is the most commonly used shielding gas.
Typical applications High-quality welds in metals such as aluminium, stainless steels, Nimonic alloys, and copper in chemical plant; sheet work in aircraft engines and structures

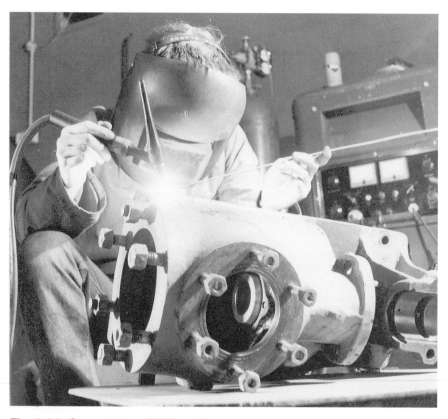

Fig. 2.26 *General view of TIG welding*

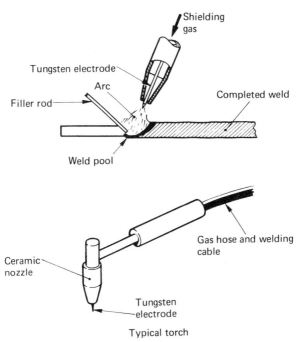

Fig. 2.27 *Schematic diagram of TIG welding*

Fig 2.28 *Equipment for TIG welding*

Manual metal arc (MMA) welding (figs 2.29 to 2.31)
Alternative names Stick-electrode welding; electric arc welding; shielded metal arc welding (in USA)
Type of operation Manual
Heat source Arc
Shielding Principally flux; some gas generated by flux
Current range 25 to 350 A
Heat input 0.5 to 11 kJ/s
Mode of operation The welder establishes an arc between the end of the electrode and the parent metal at the joint line. The arc melts the parent metal and the electrode to form a weld pool which is protected by the molten flux layer and gas generated by the flux covering of the electrode. The welder moves the electrode towards the weld pool to keep the arc gap at a constant length. The current is controlled by the power-supply unit. Electrodes are normally 460 mm long. When the electrode has been melted to a length of about 50 mm, the arc is extinguished. The solidified slag or flux is removed from the surface and the weld is continued with a fresh electrode.
Typical applications Fabrication of pressure vessels, ships, structural steelwork; joints in pipework and pipelines; construction and repair of machine plant.

Fig 2.29 *General view of MMA welding*

Fig. 2.30 *Schematic diagram of MMA welding*

Fig. 2.31 *Equipment for MMA welding*

Metal inert-gas (MIG) welding (figs 2.32 to 2 .34)

Alternative names Metal active-gas (MAG) welding; metal arc gas-shielded (MAGS) welding; gas metal arc welding (GMAW); semi-automatic welding; CO_2 welding

Type of operation Manual, but can be used with a mechanised traversing system

Heat source Arc

Shielding Gas, which must not react with the metal being welded

Current range 60 to 500 A

Heat input 1 to 25 kJ/s

Mode of operation An arc is established between the end of the electrode and the parent metal at the joint line. The electrode is fed at a constant speed by a governed motor. The electrode feed rate determines the current. The arc length is controlled by the power-supply unit, and the welder is required to keep the nozzle at a fixed height above the weld pool (usually about 20 mm).The arc area and the weld metal are protected by a gas which is chosen to suit the metal being welded. Gases commonly used are argon, argon mixed with 5% oxygen or 20% carbon dioxide, and pure carbon dioxide.

Typical applications Medium-gauge fabrications such as earth-moving equipment, plate, and box girders; sheet-metal work for car bodies.

Fig 2.32 *General view of MIG welding*

Fig. 2.33 Schematic diagram of MIG welding

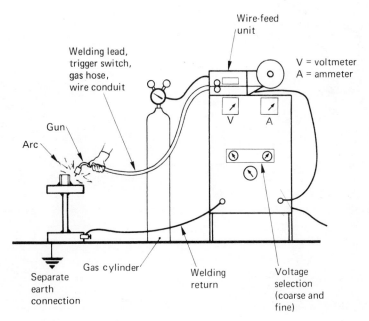

Fig. 2.34 Equipment for MIG welding

Submerged-arc (SAW) welding (figs 2.35 to 2.37)

Type of operation Mechanised
Heat source Arc
Shielding Granular flux
Current range 350 to 2000 A
Heat input 9 to 80 kJ/s

Mode of operation An arc is maintained between the end of a bare wire electrode and the parent metal. The current is controlled by the power-supply unit. As the electrode is melted, it is fed into the arc by a servo-controlled motor. This matches the electrode feed rate to the speed at which the electrode is melting, thus keeping the arc length constant. The electrode and drive assembly is moved along the joint line by a mechanised traverse system.

The arc operates under a layer of granular flux (hence 'submerged' arc). Some of the flux melts to provide a protective blanket over the weld pool. Unmelted flux is recovered and re-used.

Typical applications Joints in thick plate in pressure vessels, bridges, ships, structural work, welded pipe.

Fig. 2.35 *General view of submerged-arc welding*

Fig. 2.36 *Schematic diagram of submerged-arc welding*

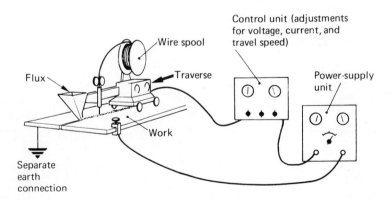

Fig. 2.37 *Equipment for submerged-arc welding (note that other types of traverse mechanism can be used)*

35

Resistance spot welding (figs 2.38 to 2.40)

Type of operation Automatic

Heat source Resistance heating at an interface

Shielding None required

Current range 100 to 50 000 A

Heat input See Chapter 12

Mode of operation The work, which is usually in the form of a lap joint, is gripped between two copper electrodes. A high current at a low voltage flows through the parent metal between the electrodes. At the interface, heat is generated by the resistance offered to the current flow. A spot or slug of metal is melted and bridges the interface. The current flows for only a short time (typically 0.06 to 3 seconds). When the current is switched off (automatically), the weld solidifies under pressure.

Typical applications Light fabrications from pressed sheet, such as car bodies and domestic washing machines. Also used for high-quality work in aircraft engines.

Fig. 2.38 *General view of resistance spot welding*

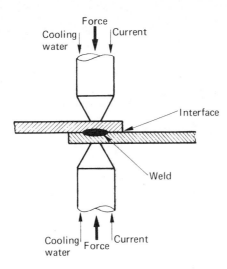

Fig. 2.39 *Schematic diagram of resistance spot welding*

Fig. 2.40 *Principal components of a resistance spot welder*
C – on/off contactor; T – timer; F – foot switch to initiate weld cycle;
Tr – transformer; L – flexible leads connecting transformer to electrodes;
I – ignitrons or thyristors to control current

3 Essential parameters in fusion welding

3.1 Parent-metal fusion

We will now consider fusion welding in more detail, beginning with the factors which influence the way the heat source melts the parent metal, since control of the weld pool is the key to making a weld of good quality. Our interpretation of what is meant by the expression 'quality of the finished weld' is very closely related to the type of service for which the joint is required, and we will be discussing this aspect at length in Chapter 8; for the moment it is sufficient to emphasise that for a weld to be acceptable there must be good bonding between the weld metal and the plate material. This can occur only when the surface of the parent metal has been melted before the added electrode or filler metal is allowed to flow into the joint. If this sequence is not followed, there will be a lack of fusion at the boundary of the weld metal, since the molten filler will be chilled by the unmelted parent metal and will freeze without having provided enough heat to melt the joint surfaces. An examination of an etched macrosection of a weld made in this way (fig. 3.1) shows up the lack of fusion.

Fig. 3.1 *Lack of side fusion in a butt weld (scale numbered in cm)*

How, then, can we avoid this situation in practice? In the case of oxy-acetylene and TIG welding, the problem is relatively straightforward since the filler metal is added separately. The first task, therefore, is to ensure that the edges of the joint have been melted and blended into the weld pool before the filler-metal additions are made. This melting must be achieved as quickly as possible, and its extent must be limited if the welder is to keep control of the weld pool. The filler metal is then added to the leading edge of the pool while the heat source is used to supply sufficient heat to the joint to prevent premature solidification.

With the manual consumable-electrode processes – i.e. MMA and MIG welding – molten metal is being transferred across the arc throughout the whole of the welding operation, and the welder must manipulate or direct the arc so that fusion of the joint faces is obtained. This is helped by the use of an electro-magnetic force which acts along the centre line of the arc to control the flow of metal in the weld pool. In practice, therefore, the welder pays attention to maintaining the correct angle between the electrode and the surface of the weld, since this influences both the distribution of the heat from the arc and the flow of the weld metal over the fused faces.

3.2 Heat input

The success of the welding operation also depends on the heat input to the joint. As soon as heat is supplied to the area to be welded, either by an arc or by a fuel-gas flame, it will start to flow away into the metal on either side, as this is at a lower temperature. This means that, if we are to achieve melting, the rate at which heat is being supplied to the joint must be greater than the rate at which it is flowing into the parent metal.

It follows that the thermal conductivity of the parent plate is probably one of the most important considerations when choosing welding conditions. The various metals used in fabrication conduct heat at different rates. Specific values can be attributed to the thermal conductivities of typical metals, but it is easier to compare them by using an arbitrary scale. Pure copper has a very high thermal conductivity and, by giving it a value of l00, Table 3.1 can be drawn up.

Table 3.1 *Relative thermal conductivities*

Metal	Relative thermal conductivity
Copper	100
Aluminium	62
Steel	14
Lead	8

During a welding operation, heat will be conducted away from the area much faster with copper than with steel. This means that, if we are to achieve fusion, heat must be supplied to a copper joint at a greater rate compared with steel of a similar thickness. On some occasions this may not be practicable, since it may call for a heat input which is in excess of that which can be provided by the process we are using. This situation is affected by the melting point of the metal.

Although aluminium has a high thermal conductivity, it is not too difficult to produce fusion as the melting point is only about 600°C, compared with 1053°C for copper, but it may be difficult to restrict the width of the melted zone.

Heating the parent metal before welding – i.e. preheating – can assist in reaching the melting point rapidly, by reducing the temperature difference between the weld and the plate. At the same time, the thermal conductivity of the metal falls with increasing temperature. Thus by heating the plate the thermal conductivity is lowered, and this in turn reduces the rate of flow of heat.

Finally, we must consider the importance of the cross-sectional area of the conductor. Normally in welding we are concerned more with thickness than with cross-sectional area, and, in general, the thicker the component being welded the faster the heat is conducted away from the joint line. We must also recognise that only rarely do we weld just one piece of metal. Usually the joint contains two or more members, each providing a path along which heat can flow; for example, a T joint has three possible heat paths – the vertical leg and both directions in the horizontal member – and so cools faster than a butt joint, which has two.

To summarise, the parameters involved in effective melting of the parent metal during welding are:

a) metal thickness, and joint type
b) thermal conductivity
c) temperature of the parent metal before welding
d) melting point
e) electrode angles and manipulation
f) heat input.

3.3 Measuring heat input in arc welding

When considering heating effects, it is usual to express the amount of heat energy in joules. The joule is the standard unit of energy. The quantity of heat energy it represents is quite small. It requires about 690 000 joules to raise one kilogram of steel from room temperature to 1500°C, i.e. just below its melting point. In welding it is difficult to measure heat input in terms of joules directly. However, we know that one watt of electrical energy equals one joule per second. So, by measuring the arc voltage and current the power input can be calculated.

Power is the rate at which energy is being supplied. When dealing with a welding arc, almost all the electrical energy is converted to heat – only a small proportion is used in the generation of the bright light and ultra-violet radiation given off by the arc. We can determine the power input, in watts, as we can readily measure current and voltage using meters or recorders:

arc power input (watts) = arc voltage × arc current.

For example, with an MMA electrode operating at 150 A, 25 V,

power input = 25 V × 150 A = 3750 watts = 3.75 kW

This value can be converted into energy:

1 watt = 1 joule per second

hence, in the example above, the energy input to the arc is 3750 joules per second (J/s).

We must remember, however, that this is the energy input to the arc. In other words, it tells us that, as long as the welder is using the arc, heat is being generated at a rate of 3.75 J/s. If the arc was kept stationary this would be the heat input to the weld at one point, but in practice the welder moves along the joint line. The heat is therefore distributed, and the input at any given point depends on the travel speed. Knowing the energy being supplied to the arc and the travel speed, it is possible to calculate the heat input and to express it as the amount of heat per unit length of weld run (i.e. in joules/mm).

Suppose a welder is using the TIG process to butt weld some stainless steel sheet which has a thickness of 1.6 mm. The current would be about 60 A and with a 2 mm arc length the arc voltage in argon shielding gas would be 14 V. The travel speed would depend on the actual component, but let us take 120 mm/min as typical. The arc power would be:

$$60 \text{ A} \times 14 \text{ V} = 840 \text{ watts}$$

$$= 840 \text{ J/s}$$

The welder moves the torch along the joint at the speed of 120 mm/min, hence:

in 1 minute the distance moved = 120 mm

in 1 second the distance moved = $\dfrac{120}{60}$ mm = 2 mm

The heat input is therefore 840 joules every 2 mm, i.e. 420 J/mm.
The formula for calculating heat input in any arc process is:

$$\text{heat input (joules/mm)} = \frac{\text{current (A)} \times \text{arc voltage (V)} \times 60}{\text{travel speed (mm/min)}}$$

With most arc processes operating at currents ranging from 100 A up to 600 A, it is more usual to quote heat inputs in kilojoules (1 kJ/mm = 1000 J/mm). The formula then becomes:

$$\text{heat input (kJ/mm)} = \frac{\text{current (A)} \times \text{arc voltage (V)} \times 60}{\text{travel speed (mm/min)} \times 1000}$$

3.4 Heat input in oxy-acetylene welding

By contrast with arc welding, in which electrical energy is converted to heat at the surface of the pool, oxy-acetylene welding depends on a chemical reaction to generate heat which is then transferred to the work.

The oxy-acetylene flame used for welding consists of two zones: an inner cone which is well defined and an outer envelope which is more diffuse. These two zones correspond to the two stages of combustion which occur in the flame (fig.3.2). As the mixed acetylene and oxygen emerge from the nozzle, the acetylene (chemical formula C_2H_2) burns to give carbon monoxide and hydrogen:

$$C_2H_2 + O_2 \longrightarrow 2CO + H_2 + heat$$

Carbon monoxide and hydrogen are both combustible gases and they in turn burn in the outer envelope when they are mixed with oxygen from the surrounding air:

$$4CO + 2H_2 + 3O_2 \longrightarrow 4CO_2 + 2H_2O + heat$$

The total quantity of heat produced by these reactions depends on the amount of acetylene which is burnt. If we require more heat, the flow rate of acetylene is increased and the oxygen supply is adjusted to give the correct type of flame. This involves a change in the size of the orifice through which the mixed gases flow, and the nozzles are graded accordingly.

The total amount of heat available is about 55 kJ per litre of acetylene. Unfortunately, not all of this heat can be used directly to produce fusion of the joint faces. The highest temperature in the oxy-acetylene welding flame is at the tip of the inner cone (fig. 3.2), and, for the most efficient operation of the system, the flame should be positioned so that this point is just above the surface of the weld pool. At this point in the flame, the only heat available for melting is that developed by the first reaction:

$$C_2H_2 + O_2 \longrightarrow 2CO + H_2$$

This represents approximately 35% of the total heat produced, and for each litre of acetylene burnt only about 19 kJ of the heat produced is used to fuse the parent metal. The heat from the second-stage reaction is not wasted, however, since the outer envelope of the flame covers the surface of the parent metal around the weld and effectively preheats the area.

Although the temperature of the acetylene flame is high enough for welding, and sufficient heat can be supplied, the rates of travel are slower than with arc welding. Briefly, there are two reasons for this. Firstly, the rate at which heat is transferred from the heat source to the work depends on the difference between the melting point of the metal and the temperature of the source. In oxy-acetylene welding the differential is not very large, and heat transfer is relatively slow. This is why it is most important to position the inner cone correctly – if there is a large gap between the tip of the cone and the surface of the weld pool, heat is transferred from a region of lower temperature and the process will be less efficient. Secondly, in arc welding much of the heat used for melting is generated at the surface of the parent metal itself and hence does not need to be transferred.

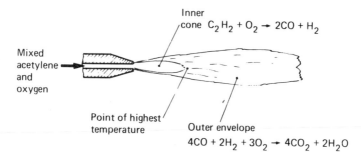

Fig. 3.2 *Simplified structure of an oxy-acetylene flame*

3.5 Summary of the main factors governing heat input

When summarising the factors governing heat input, we must therefore distinguish between arc welding and gas welding. In arc welding, the main parameters are:

- arc current
- arc voltage
- and, travel speed

In oxygen-fuel-gas welding, the main parameters are:

- fuel-gas flow rate
- position of inner core relative to the weld pool surface, and
- travel speed.

3.6 Penetration and dilution in arc welding

So far, little has been said about the way in which the parent metal is melted in arc welding. In essence, an arc is the passage of electricity across a gap between two electrodes. For this to be possible, the atmosphere in the gap must be ionised – this means that the atoms of gas have lost a free electron and are positively charged. The electrons travel from the cathode (negative) to the anode (positive), giving the flow of current. Some electrical energy is converted to heat in the arc column (fig.3.3) and is radiated into the surrounding air, but the majority is released at the end of the electrode and at the plate surface.

The proportion of the total energy supplied to the arc which is used to melt the plate will depend on whether the parent metal is an anode or a cathode, i.e. whether it is connected to the positive or negative pole of the power-supply unit. The choice of polarity is governed by a number of considerations such as arc stability, rate of electrode melting, transfer of melted metal from the electrode to the weld pool, and removal of oxides from the surface of the parent metal. These vary from one process to another, and it is not possible to generalise. TIG welding presents a special case as metal is not transferred across the arc. Using a tungsten

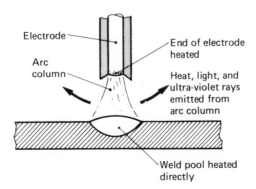

Fig. 3.3 *Distribution of heat in a welding arc*

electrode in an inert gas shield, about one third of the heat is developed at the cathode and two thirds at the anode. TIG welding is normally operated with the electrode connected to the negative side of the power supply unit. Most heat is thus developed in the metal of the joint. It must be emphasised that this distribution does not apply in welding where the electrode is melted and transferred across the arc.

The depth to which the parent metal has been fused is usually referred to as the penetration. This may be seen simply as melting into the plate, as in the case of fillet-welded 'T' joints, or as the extent to which the abutting edges in a joint have melted (fig. 3.4).

Since both voltage and current contribute to heating in the arc, a change in either will have an effect on the penetration. During welding, however, the magnitude of the voltage is usually decided by other considerations. The width of the weld and the surface profile are dependent on the arc length, which in turn is closely related to voltage. In general terms, high voltages give long arc lengths, wide flat welds, and the risk of arc instability. At the same time, oxygen and nitrogen may be drawn into the arc column. On the other hand, very low voltages (i.e. short arcs) can result in narrow welds having a high profile with the possibility of lack of fusion at the edge of the weld. The welder, therefore, chooses a voltage and arc length which give a stable arc and a satisfactory weld surface profile. Once the voltage has been fixed, current becomes the main factor in the control of penetration.

An important aspect of penetration in arc welding is the effect it has on weld-metal composition. The molten pool is made up of a mixture of electrode and parent metal. The proportions will be determined to a large extent by the heat

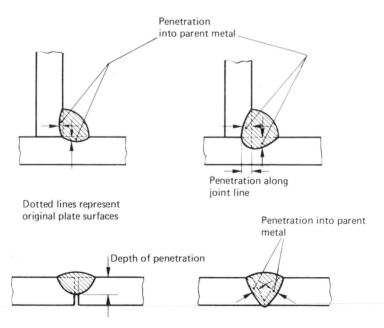

Fig. 3.4 *Penetration in T and butt joints*

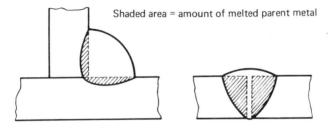

Shaded area = amount of melted parent metal

Fig. 3.5 *Effect of dilution on composition*

input, since this controls both the rate of electrode melting and the penetration. It is useful to view the weld metal as metal deposited from the filler wire or electrode diluted by mixing with the melted material from the workpiece. By convention, dilution is expressed as the percentage of melted parent metal in the weld (fig. 3.5) and can be used to predict individual alloy content in the molten pool.

$$\text{Dilution} = \frac{\text{shaded area}}{\text{total area of weld}} \times 100\%$$

$$c_w = c + \frac{D}{100}(c_p - c_f)$$

where c_w = percentage of a given element in the weld
c_f = percentage of this element in the filler or electrode
c_p = percentage of this element in the parent metal
D = percentage dilution

This concept is of particular value in designing procedures to avoid cracking when fabricating some stainless steels and aluminium alloys – the composition must be carefully controlled to ensure that the weld metal has adequate strength at solidification. Again, when coating the edge of a carbon-steel blade with wear-resistant weld metal, the properties of the deposited layer may be impaired if too much steel is melted into it. In such cases, if the acceptable dilution can be determined, it is possible to specify the maximum allowable penetration.

3.7 Position of welding

Ideally, the work should be positioned during welding so that the molten weld metal is held in place by gravity. This is called the flat position (fig. 3.6) and gives the welder the most favourable conditions for controlling the weld pool. It also enables high currents to be used, leading to faster welding. This implies that the work can be turned or manoeuvred. Many fabrications do not lend themselves to this treatment, and much of the welding in industry is done 'in position'.

Three main positions, in addition to flat, can be identified: horizontal, vertical and overhead. There is a subdivision of the horizontal position, known as the horizontal-vertical, which relates specifically to T joints in which the axis of one member is vertical while the other is horizontal. In all these positions the metal

45

Fig. 3.6 Positions of welding

tries to run out of the joint under the effect of gravity, and the welder's technique must be modified to combat this tendency. The welder controls the weld by lowering the heat input to reduce the fluidity and to give a small pool which solidifies before it has time to run out of the joint. At the same time, the direction of the arc, i.e. the angle between the electrode and the weld surface, can be varied to position the weld pool to the best advantage (Table 3.2). The arc force also helps to keep the weld pool in place.

46

	ELECTRODE ANGLE (A)		
Welding position	Flat	Horizontal-Vertical	Vertical
MIG/MAG Welding	75° – 80°	70° – 75°	65° – 70°
MMA Welding	70° – 75°	65° – 75°	70° – 80°
TIG* Welding	70° – 80°	70° – 80°	85° – 90°

45° 45°   45°-50°   45° 45°   A   A   A

*In TIG Welding the filler wire is introduced into the leading edge of the weld pool at an angle of 15° – 20° to the pool surface.

Table 3.2 *Electrode angles for manual arc welding*

The maximum current is lower in positional welding. Whereas 350 A can be readily used for joints in the flat position, the welder would have considerable difficulty in working above 160 A when welding overhead. It follows that the sequence used to deposit a weld of a given size differs from one position to another. This can be illustrated by considering the deposition of a typical fillet weld. The size of a fillet weld can be specified in a number of ways. This aspect is discussed in greater detail in Chapter 7, but for our present purpose we will specify a leg length of 10 mm. This is the distance from the root of the weld to the toe and is, in effect, one side of the triangle formed by the fillet (fig. 3.7). In the flat position, with a current of 300 A the weld could be deposited in one pass. The welder moves the arc from side to side to ensure that the faces of the joint are properly fused. This is known as weaving and it produces a smooth flat surface. Using this technique, a one-metre length of weld could be deposited in about 10 minutes.

If the joint is in the horizontal-vertical position, attempts to deposit the weld in one pass result in a large uncontrollable pool which tries to flow on to the surface of the horizontal plate. The resultant weld is misshapen and there is overlap or lack of fusion at the toe. To obtain a uniform shape, the weld is deposited in three runs, using a lower current of about 200 A. Although the travel speed for each individual run is faster, the total time for the joint is 15 minutes, i.e. 50% longer than in the flat position.

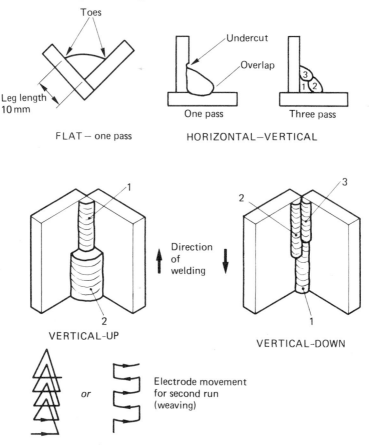

Fig. 3.7 *Deposition of 10 mm-leg-length fillet weld*

When the joint is positioned so that the weld is deposited overhead, the current is reduced to 160 A and four or five runs may be needed to obtain the required size. The time has now increased to about 24 minutes.

When the plates are arranged so that the joint line is vertical, the weld can be started at the bottom and the arc moved upwards. This is the conventional method of fillet welding a joint in this position and is often referred to as the vertical-up technique. Two runs are required at about 145 A and the welder weaves the electrode along a triangular path to achieve fusion of the joint faces. The weaving also helps to distribute the heat and control the weld-pool fluidity. If the welder experiences difficulty in preventing the surface from sagging at the centre-line, the current is reduced to about 120 A and the weld is deposited in three passes using a smaller weave. The weld times range from 20 to 25 minutes, depending on the current used.

Alternatively, a vertical-down technique can be used. Welding starts at the top of the joint and the electrode is pointed upwards at the deposited weld metal. The rate of travel is critical, as the molten metal must not run down the joint ahead of the arc – if it does, fusion of the parent metal may not take place. The weld pool must be relatively small, and there is little scope for weaving. This means that our

Fig. 3.8 *Square-edge butt-welded joints*

10 mm-leg-length fillet weld may need five or even six weld runs. On the other hand, travel speeds with the vertical-down technique are comparatively high, and the overall time for the joint is about 17 min/m.

3.8 Penetration and edge preparation

The thickness of metal which can be butt-welded using square edges for the joint depends on the amount of penetration available. In arc welding, the limiting factor is current. With the MMA and TIG process the maximum thickness is 3 mm if the weld is made from only one side or 5 mm if a second run can be deposited on the reverse side (fig. 3.8). The same limitations apply to MIG welding at currents below 200 A; but with high currents, say 400 to 450 A, plates having a thickness of 6 mm can be butt-welded from one side with square edges. With thicker material, the edges must be cut back to provide access for the arc. The simplest edge preparation (fig. 3.9) is a single bevel. When the joint is assembled, a groove is formed which is filled by depositing a number of weld runs into it. Each run is fused into both the surface of the preceding weld and the side walls of the groove, to ensure bonding through the complete thickness of the joint. The first weld or pass to be deposited is known as the root run (fig.3.10). It often requires considerable skill to produce a satisfactory root run, since the

Fig. 3.9 *Single-V preparation*

1   =   root run
2,3,4   =   filling runs
5   =   final or capping run

Fig. 3.10 *Typical welding sequence for a single-V butt weld*

welder must fuse the root faces but at the same time keep control of the width of the weld bead as the molten metal is unsupported on the underside. The risk of producing either excessive penetration or lack of fusion in a root run is always very great, and in some areas of work it is common practice to remove the root run by gouging a groove along the line of penetration after the 'V' has been filled. The joint is then completed by depositing a sealing run from the reverse side.

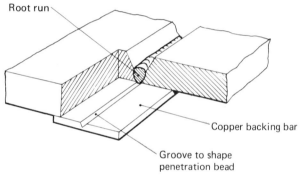

(a)   Temporary backing bar, with groove, to
      support root run (removed after welding)

(b)   Permanent backing strip

Fig. 3.11 *Temporary and permanent backing for butt welds*

The root gap plays an important part in controlling penetration in the root of a joint. The size of the gap depends on the process and the application, i.e. plate or pipe, type of joint, and position of welding. It is most important that the gap is constant – if it varies along the joint, the welder will have great difficulty in achieving consistent penetration. Where it is not possible to ensure uniformity of fit-up or adequately to control root fusion, some form of backing can be used. This may be temporary, in that it is removed after welding, or the root run may penetrate into a strip of metal which becomes a permanent part of the joint (fig. 3.11).

Various types of edge preparation can be used (fig. 3.12), and the choice of the

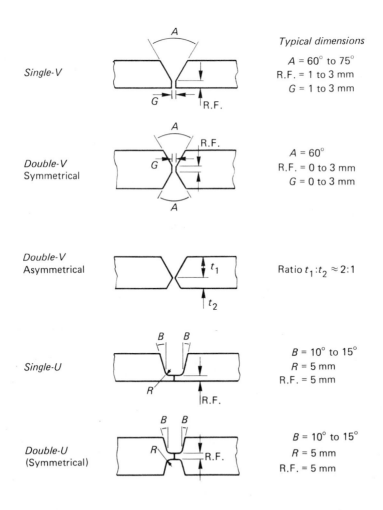

Fig. 3.12 *Typical edge preparations for butt joints*

51

most suitable is influenced by a number of factors. Some of these, not necessarily in order of importance, are:

a) type of process
b) type of work
c) position of welding
d) access for arc and electrode
e) volume of deposited weld metal
f) dilution
g) cost of preparing edges
h) shrinkage and distortion.

4 Control of welding parameters

4.1 The importance of control

In Chapter 3 we developed the idea that for a weld to be considered of acceptable quality there must be bonding between the weld metal and parent material, but it would be pointless if this occurred only at intervals along the joint. If the weld is to have full strength, there must be bonding at all points throughout the total length of the joint, which means that the heat input must be controlled to ensure it is always above the minimum level required for the thickness being welded. It is then left to the welder to use the correct manipulative technique and travel speed to achieve continuity.

Having achieved fusion, the welder must also aim to keep the size of the weld constant within certain limits and to produce a uniform profile or shape. Consistency is important both to the designer, who will specify a weld size to suit the load which the joint is to carry, and to the estimator, who will want to know that the time to deposit a weld will remain reasonably reproducible .These considerations pose the question: how do we control the heat input? The answer to this obviously depends on the nature of the heat source, and for the two types which we have considered so far we need to analyse the main parameters governing input.

4.2 Control in oxy-acetylene welding

In the case of oxy-acetylene welding, the analysis is relatively straightforward . The heat used to produce melting is provided by the combustion of acetylene at the orifice of the nozzle. The more acetylene supplied, the greater will be the heat generated, which means that the flow rate of acetylene must be controllable. When the oxy-acetylene flame is used for welding, the heat input to the joint will also depend on the efficiency of the combustion. The maximum total heat is produced when the acetylene is completely burnt in an oxidising flame, i.e. a flame containing an excess of oxygen over that needed to combine with the acetylene, but this setting is rarely employed for welding since it does not give the hottest flame and can cause oxidation of the weld metal. More normally a ratio of acetylene to oxygen is chosen which produces a neutral flame – i.e. one in which there is no excess of either gas. The relative amounts of acetylene and oxygen are adjusted by means of flow valves built into the torch. The gas supplied to the nozzle is therefore a controlled mixture of oxygen and acetylene which must be measured and regulated.

The use of this principle can be seen in a simplified layout of a typical oxy-acetylene system (fig. 4.1). The regulators which are attached to the outlets from the cylinders not only reduce the pressure of the gas from that of the cylinders to

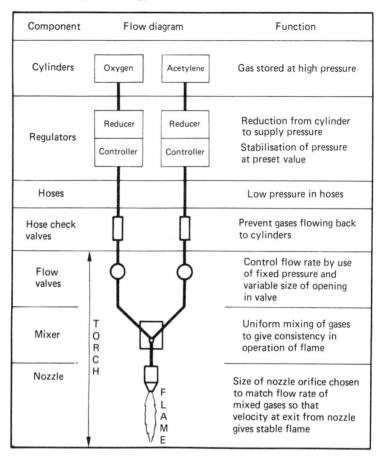

Component	Flow diagram	Function
Cylinders	Oxygen Acetylene	Gas stored at high pressure
Regulators	Reducer Reducer	Reduction from cylinder to supply pressure
	Controller Controller	Stabilisation of pressure at preset value
Hoses		Low pressure in hoses
Hose check valves		Prevent gases flowing back to cylinders
Flow valves		Control flow rate by use of fixed pressure and variable size of opening in valve
Mixer	T O R C H	Uniform mixing of gases to give consistency in operation of flame
Nozzle	F L A M E	Size of nozzle orifice chosen to match flow rate of mixed gases so that velocity at exit from nozzle gives stable flame

Fig. 4.1 Schematic layout of oxy-acetylene system

the level required to achieve correct flow rate and flame stability, but also keep the reduced pressure constant, thus ensuring uniformity of operation. The flow valves in the torch alter the amounts of acetylene and oxygen being supplied to the mixer. These must be adjusted in conjunction with the regulators, since the rate of flow depends on both the pressure at the entry to the valve and the size of the valve opening. The mixed gases are fed to the nozzle, where the diameter of the orifice must be chosen to match the volume of the acetylene and oxygen mixture being used, so that the velocity of the gas issuing from the nozzle enables a stable flame to be maintained. A sectional view of a modern oxy-acetylene welding torch (fig. 4.2) shows how these features are incorporated.

4.3 Control in arc welding

Turning now to the use of an arc for welding, we can summarise its main features:

a) A welding arc is the passage of electricity across a gap between an electrode and the parent metal or work.

b) Heat is generated at the surfaces of the electrode and of the parent metal.

54

Oxygen

Acetylene

To nozzle

Mixing chamber

Body

Oxygen valve

Hoses to cylinders

Mixed gases

Nozzle

Acetylene valve

Hose check valves to prevent flow of gas back towards cylinder

Fig. 4.2 *Diagrammatic sections of a typical oxy-acetylene welding torch*

c) Heating at the electrode can cause melting and the transfer of metal to the weld pool.

d) Heating at the parent metal produces fusion of the joint faces.

e) Heat input to the weld is a function of arc voltage, arc current, and travel speed.

f) Arc length is related to arc voltage.

The arc characteristic

As with the oxy-acetylene flame, in arc welding the welder adjusts the travel speed to give a uniform weld, and the requirement is for an arc with a constant voltage and current. When the arc is operating in a stable manner, the voltage and current are related . The relationships can be shown graphically by plotting the appropriate value of arc voltage for various currents (fig. 4.3). The resultant curve is called the *arc characteristic*.

There are three distinct sections on the curve. First at low currents as the current increases the arc voltage falls. Arcs operating in this region tend to be unstable and special techniques are employed to enable welding to be done at low heat inputs. Secondly, there is a minimum where increases in current over a short range produce little change in voltage. Finally, the voltage rises with current – this is the part of the arc characteristic which covers the normal welding range.

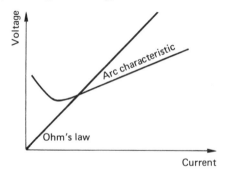

Fig. 4.3 *Typical arc characteristic compared with Ohm's law*

It is of interest to compare the typical arc characteristic with the relationship that would be predicted by Ohms Law. This states that the current flowing through a resistor is proportional to the voltage applied across it. On a graph there would be a straight line relationship between voltage and current, and the line would go through the origin (fig. 4.3). Even in the range where the voltage is rising, the characteristic shows a slight curve confirming that the arc does not follow Ohms Law and is not purely resistive. The slope of the arc characteristic above the minimum is determined principally by the metals involved and the atmosphere in the arc gap. Arc length has only a minor influence on the slope.

4.4 Arc-length control in TIG welding

Typical characteristics for four arc lengths between a tungsten electrode and copper in an argon atmosphere are shown in fig. 4.4. From these we can read off

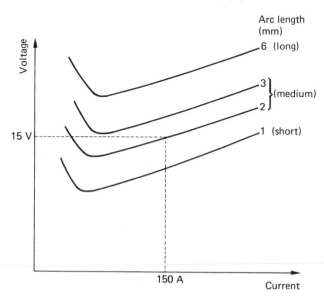

Fig. 4.4 *Arc characteristics for welding copper in an argon atmosphere (TIG welding)*

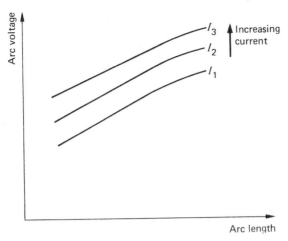

Fig. 4.5 *Relationship between arc voltage and arc length*

the values of voltage at various currents for different arc lengths and plot the results (fig. 4.5) to show that, for a particular current, the voltage increases as the arc is lengthened. This observation is of considerable importance in solving the practical problem of controlling the arc.

For a start, let us take the manual welding of copper sheet using the TIG process and suppose that the operation requires 150 A with an arc length of 2 mm. The power-supply unit is set to give an output of 150 A. The operator strikes the arc, establishes a weld pool, and travels along the joint. The arc characteristics in fig. 4.4 tell us that at 150 A for a 2 mm arc to be operating satisfactorily the voltage should be 15 V. This value will be maintained as long as the power-supply unit continues to deliver 150 A and the welder keeps the arc length at 2 mm.

In reality many things may cause the arc length to vary, with the result that the voltage rises or falls in sympathy and the operating point moves from one characteristic to another. If the power supply is able to provide only one voltage at the selected current, the arc would be forced to work at an incorrect voltage and it would become difficult for the operator to control the melting in the joint, i.e. the arc would be unstable. For stable operation the power-supply unit must allow the voltage to vary while keeping the current substantially constant. It follows from this that the output of the power-supply unit must be designed to meet the specific requirements of the TIG welding process.

It is usual to describe the output in terms of the voltage and current relationships for various output settings. A typical output characteristic for a TIG power unit is shown in (fig. 4.6). With no load across the output, the power-supply unit is not delivering current and the voltage is higher than it is during welding. This is the *open circuit voltage* (o.c.v.) and is usually in the range 50 to 80 V. Once the arc is drawing current from the power supply, the voltage falls and over the operating range of 10 to 30 V the current varies to only a small extent.

Power-supply equipments with this type of output are known as 'drooping' characteristic' units, and a series of output curves is produced by changing the current selector setting (fig. 4.7). Sometimes such units are referred to as constant

Fig. 4.6 *Typical output characteristic for a power-supply unit used in TIG welding*

current machines, but this is not strictly a correct description since the output shows some change of current with voltage.

If the arc and power-supply-output characteristics are plotted on the same graph, their intersection gives the working voltage and current. Taking the example given above for the welding of copper, the output for a 150 A setting on the power-supply unit is shown in fig. 4 8, together with the arc characteristics for arc lengths of 1, 2, and 3 mm. With a 2 mm arc the inter section is at 15 V for 150 A (point A), but if the welder increases the arc length to 3 mm the operating point now moves to the intersection of the 3 mm arc characteristic and the 150 A output line, i.e. 16.5 V, and the current level falls to about 143 A (point B). Conversely, if the welder shortens the arc length to say 1 mm, the operating point moves to point C and the voltage falls to 133 V with a corresponding current of 156 A. These results can be summarised:

Arc length (mm)	Voltage (volts)	Current (amperes)	Power input (watts)	Variation in power (%)
1	13.3	156	2074	−7.8
2	15.0	150	2250	0
3	16.5	143	2359	+ 4.8

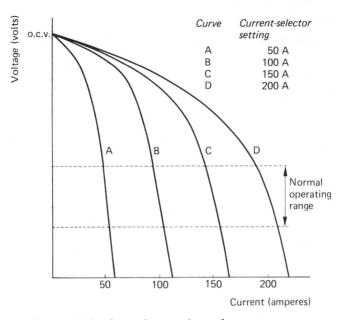

Fig. 4.7 *Output characteristics for various settings of current*

$X = 143$ A
$Y = 150$ A
$Z = 156$ A

Fig. 4.8 *Variations in voltage and current with change in arc length (TIG welding)*

The important point to recognise is that, as the welder makes small changes in the arc length, possibly as a result of involuntary hand movements and so on, the power input to the arc remains within ± 8% of the preset value, which is much better than the accuracy to which most welders can control travel speed. It might be concluded from this data that by increasing the arc length the welder would increase the amount of heat in the weld. This is not true. The values quoted are for power delivered to the arc. Increasing the arc length means that a higher voltage is needed to overcome the increased resistance to the flow of the current resulting from the longer distance it has to travel. The extra power supplied is in effect used up in the arc itself and does not significantly increase the amount of heat developed in the weld pool. At the same time, because the arc tends to be cone shaped, the heat is distributed over a larger area of the joint.

The only way to increase the energy delivered to the weld pool simply by increasing the voltage at a given current is to change the atmosphere in which the arc operates. This is true for all arc welding processes. In TIG and MIG welding this can be done by changing the shielding gas (p92 & p95) whereas for MMA welding it would mean a change of electrode type as the arc atmosphere depends on the flux covering (page 79). It is normally more practicable to alter current if we wish to increase the heat input.

4.5 Arc-length control in MMA welding

The principles discussed above can be applied to MMA welding. The arc characteristic has a similar shape but different values for the voltage/current relationship. The behaviour of the arc in MMA welding is complicated, however, by the melting of the electrode. If the electrode is held stationary, the arc length increases progressively as more of the electrode is melted off. Therefore, the electrode holder must be moved towards the weld at the same rate as the electrode is shortening. This latter is known as the *burn-off* rate, and the volume of metal melted each second is dependent principally on the current – the higher the current the faster the burn-off rate, and vice-versa. In MMA welding, the consistency of the weld is therefore almost entirely dependent on the welder's skill in estimating the arc length and manually adjusting the electrode feed rate.

4.6 Self-adjusting arcs in MIG welding

With MIG welding the situation is in a sense the reverse of that described for MMA welding above, since the voltage setting of the power-supply unit, not the welder, governs the arc length. How then is the arc contoured?

First we must recognise that with the smaller diameter wires used in the MIG system the burn-off rates are far higher than in MMA welding and they vary much more with current. With only a small fluctuation in current, a significant change in burn-off rate is produced. Typical burn-off curves for low-carbon-steel wires in carbon-dioxide shielding are given in fig. 4.9.

An important point to note is the effect of electrode diameter. At 200 A the burn-off rates are 2.5, 5.1, and 10.4 m/min for 1.6, 1.2, and 0.8 mm diameters respectively. Equally significant is the change in burn-off rate produced by a given change in current – at 220 A the rates are 2.8, 5.6, and 11.5 m/min. Used in

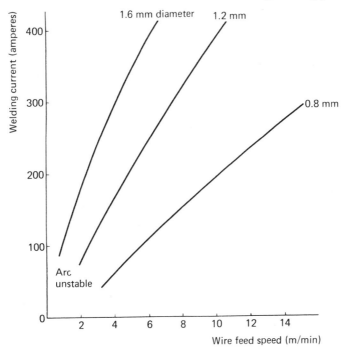

Fig. 4.9 *Relation between wire feed rate and current (i.e. burn-off curves) for three diameters of steel electrodes in CO₂ welding.*

conjunction with a power-supply unit which keeps the voltage reasonably constant over a wide current range, this feature of MIG welding provides the basis for the control of the arc.

A suitable output characteristic for a MIG power supply is shown in fig. 4.10. There is a straightline relationship between voltage and current, with the voltage falling by about 2 V for every 100A increase in current. This type of supply is often referred to as 'flat' or 'constant potential'.

Consider an arc operating at 300 A, 35 V (point A on fig. 4.10), when the corresponding burn-off rate would be 7.6 m/min. Suppose the arc length increases, with the result that the voltage rises to, say, point B. This is accompanied by a decrease in current, giving a correspondingly lower burn off rate (6.3 m/min). In the MIG system the wire feed speed is constant, so the wire electrode is feeding into the arc area at a faster rate than it is burning back, since the burn-off rate has been reduced by the fall in current. The tip of the electrode approaches the surface of the weld pool and the arc length shortens, which means that the voltage returns to the original value. When this has been reached, the current will have risen and the burn-off rate once again matches the wire feed rate. The system has been returned to equilibrium. Conversely, if the arc length shortens, the voltage falls, the current rises, and the bum-off rate increases. The wire, therefore, melts off faster than it is being fed into the arc area and the arc length is continuously increased until it returns to the preset value.

This sequence of events is known as self-adjustment, and the key to its effective operation lies in the magnitude of the change in burn-off rate. Suppose,

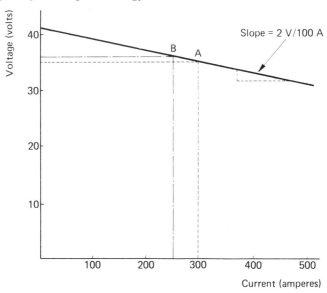

Fig. 4.10 *Output characteristics for a constant-potential power-supply unit*

once again, the arc length has been increased and the current has changed from I_1 to I_2 (fig. 4.11). Although the burn-off rate, corresponding to I_1, is now B_2, the wire is still feeding at the speed of the original burn-off rate (B_1), with the result that the tip of the wire is advancing towards the weld pool at a speed equal to ($B_1 - B_2$). Hence, as the change in burn-off rate gets larger for a given difference in current ($I_1 - I_2$), the speed with which the electrode tip returns towards its original burn-off position increases.

With electrode wires in the range 0.8 to 1.6 mm diameter, this requirement for rapid self-adjustment is readily met. Thus, for example, with a 1.2 mm diameter steel wire in carbon-dioxide shielding the change in burn-off rate produced by a 20 A change in current is 0.51 m/min, which means that a 1 mm change in arc length is corrected in about:

$$\frac{1}{510} \times 1 \text{ min, i.e. } 0.13 \text{ second.}$$

Again, with an 0.8 mm diameter wire the change in burn-off rate for 20 A change in current is 1.02 m/min, corresponding to a correction time of approximately 0.059 s per 1 mm change in arc length.

Pursuing this line of thought, we can readily see that with MMA welding the correction times would be long because the changes in burn-off rates are small. For example, a 4 mm electrode operating at 200 A melts off at about 0.25m/min and a 20 A change in current produces only 0.02 m/min change in burn-off rate. An alteration of 1 mm in the arc length would therefore need about 3 seconds to be corrected by self-adjustment, which is much longer than it would take the operator visually to recognise and correct the change. It follows that, for MMA welding, better results are obtained if the current is kept constant by the use of a drooping-characteristic power supply and the welder is left to control the arc length as described above.

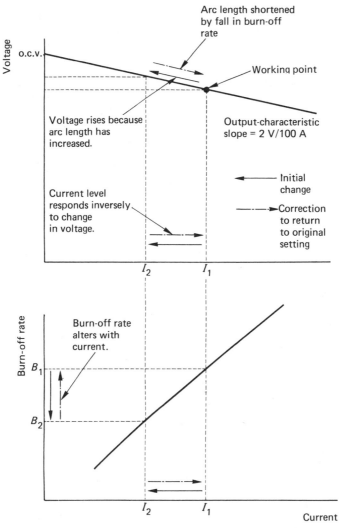

Fig. 4.11 *Changes in voltage, current, and burn-off rate with changes in arc length (MIG welding)*

4.7 Summary of heat-input control in arc welding

Summarising the control of heat input in arc welding, there are five main parameters which must be considered:

a) arc length
b) arc voltage
c) electrode feed rate
d) arc current
e) travel speed.

In the three manual arc-welding processes we have considered, the travel speed is under the direct control of the welder and the degree of consistency depends

63

critically on individual skill. The way in which the other four parameters are controlled depends on the process and can be summarised as shown in Table 4.1.

Table 4.1 *Controller of welding parameters in TIG, MMA, and MIG welding*

Process	Arc length	Voltage	Electrode feed	Current
TIG welding	Welder	Welder	Not applicable	Power supply
MMA welding	Welder	Welder via arc length	Welder	Power supply
MIG welding	Power supply via voltage	Power supply	Wire feed	Electrode speed via wire-feed motor

4.8 Supplying power for arc welding

The power required for arc welding is supplied by an electrical unit, known as a power source, which is designed to match the process. The power source in arc welding must

a) isolate the welding circuit from the mains supply
b) provide the voltages required for the welding operation
c) supply current at the selected level
d) have an output characteristic which is compatible with the arc system
e) incorporate a low-voltage supply for the operation of auxiliary units.

Electrical supply systems

In Europe, electricity is supplied via a distribution grid at high voltages (of the order of 11 000V) to reduce the size of cable required. The voltage is reduced for local distribution to users and is normally between 390V and 440V. The current is alternating with a frequency of 50 Herz (50Hz) which means that there are fifty positive and negative cycles per second; in the USA the frequency is 60Hz. For most industrial establishments, the current is carried in three cables; this is called a three-phase supply. The current in each cable is at a different point in the alternating cycle. When the current in the first cable is just changing from negative to positive (fig.4.12), that in the second cable is one-third of the way through the complete positive-negative sequence whilst in the remaining cable the current is two-thirds into the cycle. The currents are said to be 120° out-of-phase. This system is more efficient than using just one cable and provides a choice of a high voltage (390 V to 440 V) or a lower voltage for ancillary equipment (typically the 240 V used for domestic equipment such as TV sets and washing machines).

Power units in fabrication shops normally operate from a 440V mains supply. A few units are designed for 240 V operation, but they have only limited power output.

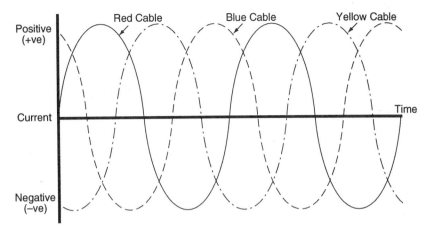

Fig. 4.12 *Three phase current supply*

Generators and alternators

When a coil of wire is rotated in a magnetic field a voltage is induced which causes a current to flow in a load connected across the coil. This principle is employed in generators and alternators used as power sources for welding. Both consist of an armature rotating in a magnetic field produced by coils which are wound onto a number of shaped steel blocks called 'poles' (fig.4.13). The output is regulated by adjusting the current flowing in these windings. The armature is rotated at a constant speed by a motor unit. An electric motor is generally the first choice where a mains supply is available. In the absence of electrical power, the welding generator can be driven by a governed petrol or diesel engine. The current is collected by carbon blocks, called brushes, which are held in contact with a copper ring (commutator) connected to the armature coil. In practice there

Fig. 4.13 *Schematic of generator*

65

is more than one armature winding and the commutator is made up of a number of sections which can be arranged to give an alternating current output (alternator) or a direct current (generator).

For many years, motor-driven generators were the most commonly used supply units for welding with d.c.. To a large extent they have been replaced in fabrication shops by rectifier units as the latter are quieter in operation and require less maintenance. Generators are still preferred for site work, where they are essential if a mains supplies is not available, and for some shop applications such as welding large diameter pipes, where independent control of open circuit voltage (OCV) and slope is required. They can also be useful when mains voltage fluctuations are troublesome.

Alternators have not been so widely used, but it has been claimed that an alternator in which the armature speed has been increased to give a higher frequency a.c. output can improve arc stability. When a.c. is required for welding, the power source is almost always a transformer connected to a 440 V supply.

Alternating current for arc welding

Alternating current (a.c.) is widely used in MMA welding of steel and TIG welding of aluminium and its alloys; it is rarely applied to MIG welding even though it is technically feasible to operate consumable gas-shielded arcs with a.c.. A typical transformer power source consists of two principal parts:

a) the transformer which reduces the mains voltage to that required for effective operation of the arc, and
b) a means of adjusting the current supplied to the arc.

The transformer has two separate coils of wire wound onto a laminated iron core (fig.4.14). The primary winding is connected to the mains supply, whilst the secondary winding provides the output to the arc. The primary and secondary circuits are electrically isolated from each other but are coupled by the magnetic field created in the core by the flow of current in the primary winding. The ratio of the number of turns in the primary and secondary windings determines the output voltage. For example, a ratio of 5.25:1 gives an OCV of 80 V from a 420 V input.

Fig. 4.14 Transformer

In the conventional welding transformer there is only one primary winding which may be connected to a single phase line, if available, but more typically it is connected across two-phases of a three-phase supply. This places an unbalanced load on the mains and in large shops it is often necessary to make special provisions to distribute the load more uniformly across the three phases.

The current is adjusted by varying the amount of inductance in the secondary circuit. Typically, an inductance is a coil of copper wire which may be wound onto a laminated iron core; when incorporated into welding equipment, it is often called a reactor.

A reactor placed in the line from the transformer to the electrode limits the level of alternating current which can flow in the welding circuit. The precise level depends on the amount of inductance in the circuit, and, for current control during welding, a means of altering this inductance must be provided. Each setting of the inductance produces a different output characteristic (fig.4.7). Two types of reactor are commonly used in power-supply units to give inductive control of current: tapped reactors and saturable reactors.

Tapped reactors

Tapped reactors have a series of connections to the windings so that sections of the reactor can be taken out of circuit as required (fig 4.15). Coarse and fine controls are provided, but only a limited number of settings can be accommodated. In an alternative design, the core of the reactor is moved in or out of the coil to vary the inductance. With multi-operator sets, where one transformer provides three or six outlets (fig.4.16), the current flowing in each welding circuit must be controlled independently and a separate reactor must be included in each lead.

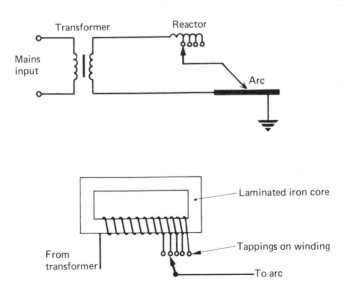

Fig. 4.15 Tapped reactor used to regulate welding current

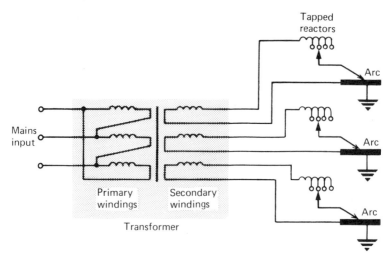

Fig. 4.16 *Multi-operator transformer unit*

Saturable reactors

Saturable reactors use an auxiliary supply of direct current to control the level of magnetism in the core of the reactor, which, in turn, governs the amount of inductance (fig.4.17). Thus, by altering the d.c. in the control winding, the welding current can be regulated. In addition to providing continuously variable adjustment, saturable reactors can be remotely controlled. They are, however, appreciably more expensive than tapped reactors and are normally only used with TIG welding.

All these designs provide good control of current and a suitable output for MMA and TIG welding; the choice is determined mainly by cost and individual preference.

Fig. 4.17 *Saturable reactor*

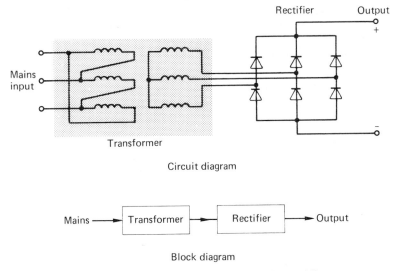

Circuit diagram

Mains → Transformer → Rectifier → Output

Block diagram

Fig. 4.18 *Simple three-phase full-wave rectifier unit for welding*

Direct-current supplies – rectifiers

In a rectifier unit, the a.c. output from a transformer is fed to a rectifier which converts it to d.c. (fig. 4.18). The current from the rectifier is not completely constant. It varies about a mean value, each variation corresponding to a half cycle of the alternating current input to the rectifier. This effect is called a *ripple* (fig. 4.19). If a single-phase input is used, the d.c. has a pronounced 100 Hz ripple and for most applications some form of smoothing may be required. With a three-phase transformer, i.e one that has three windings connected across the three-phases of the mains input, three rectifier units can be used in parallel and their outputs combined. The result is a significantly less pronounced ripple with the added

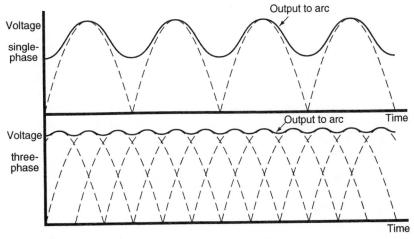

Fig. 4.19 *Ripple voltage*

69

advantage that a more uniform load is imposed on the mains supply – most conventional rectifier units for welding are designed for three-phase operation.

For MIG welding, the transformer winding is tapped so that the output voltage can be selected to suit the arc length. Since there is no requirement for current control, the power supply unit consists simply of the transformer and a rectifier.

In the case of MMA or TIG welding, a drooping output characteristic is required. This is obtained by inserting a reactor into the a.c. line between the transformer and the rectifier (fig 4.20). It should be noted that a reactor offers impedance only to a.c. – placed in the d.c. output from the rectifier, a reactor has little or no effect on the steady flow of current. It does, however, oppose any changes of current level, slowing down the rate of rise or fall. This feature is utilised in low-current MIG welding of steels (page 100).

Fig. 4.20 *Saturable reactor in d.c. unit*

By providing extra connections to the output from the reactor in a transformer rectifier set, it is possible to produce a combined a.c./d.c. unit suitable for MMA and TIG welding. This type of unit is of considerable attraction where there is a mix of work, but it costs more than the equivalent individual a.c. or d.c. unit. Combined a.c./d.c. sets usually have a single-phase input.

Operating problems with power-supply units

The power-supply units discussed so far all have the merit of being relatively simple and reliable. To a large extent they have met the requirements of industry, but they have deficiencies or drawbacks. Fortunately, these can be offset to a large extent by a combination of the skill of the operator and the tolerance of minor changes in voltage and current which is a feature of arc-welding systems. The deficiencies become more apparent and start to pose problems as we look towards higher quality, greater consistency and the use of automated plant. Five particular problems can be readily identified.

1) We have to recognise one of the most common operating problems: fluctuations in mains voltage which are reflected in the arc voltage. If the mains input voltage falls, the output voltage to the arc falls in proportion. Although it is usually assumed that the mains voltage is constant at a fixed value, under normal circumstances in a welding shop the voltage at the input to the welding

power unit can be anywhere between 390V and 440V on a three-phase supply. Several factors can affect the mains voltage, all beyond control of the welder.

From the point of view of the welding operation, the significance of these variations is that they are immediately reflected in the behaviour of the arc. With MMA and TIG welding, the welder can readily adjust the arc to offset the effect of a change in arc voltage; but the current also alters causing problems of, say, burn-through in the root run. In MIG welding, on the other hand, a drop in the input voltage can result in a shift in the working point on the static characteristic (page 62) which is not corrected by self-adjustment and continues until the mains voltage corrects itself. This could affect both the width of the weld and the penetration. It can also be the cause of a lack of fusion .

2) The current control systems which we have discussed do not have the ability to produce rapid changes. In Chapter 6 we will look at techniques such as pulsing which require the current to be changed from a high to a low level 50 to 100 times per second. By their very design, the inductive devices used in conventional units resist changes in current, and it becomes difficult to achieve rapid response.

3) It is difficult to provide remote control on most of the power units we have studied in this chapter. The exception is the use of a saturable reactor, but, as we have seen this type of unit is expensive.

4) Maintaining a pre-determined setting over a period of time is an essential requirement for quality control. In MMA and manual TIG welding this simply involves monitoring the current and relying on the welder's skill to achieve consistency. With MIG welding there is a greater number of variables, and a simple transformer-rectifier unit cannot provide the more sophisticated level of control which is often required.

5) The units tend to be large and heavy and they are not easily manoeuvrable. Most of the weight is in the transformer and inductors, if they are fitted. Other components such as rectifiers, relays and fans are bulky rather than heavy.

The development of solid-state devices, such as the transistor, has provided an opportunity to radically redesign power units and overcome the above problems.

Thyristor-controlled rectifiers

The rectifier in a conventional welding power unit is essentially a static component. It simply accepts the alternating current from the transformer or saturable reactor and converts it to d.c. without exercising any control. If a thyristor is used in place of the conventional rectifier, however, the current flow can be altered in response to a remotely controlled command signal (fig. 4.21).

A thyristor is a solid-state device which allows current to flow in one direction, i.e. it behaves as a normal rectifier. However, in the appropriate half cycle it will conduct only after being 'fired' by a voltage pulse applied to the control terminal. Current will then flow for the remainder of that half cycle, until its direction reverses and the thyristor ceases to conduct. In a typical a.c. waveform, the mean current depends on the time during which current is flowing. Thus, if the firing of the thyristor is delayed after each time the voltage changes

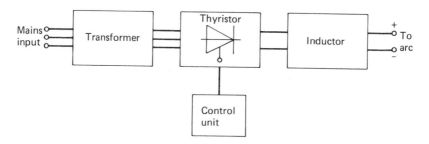

Fig. 4.21 *Thyristor-controlled welding rectifier*

from positive to negative and vice versa, the current flows for a shorter time in each half cycle and the mean current is reduced (see the diagram for the primary circuit in fig.12.7). The significance of this is that we can use the firing point of a thyristor to adjust the output from a welding power unit. In this way, thyristors can do the jobs of both the rectifier and the saturable reactor.

The main drawback of the thyristor is the very pronounced ripple in the output current, it can be reduced by including an inductor in the circuit.

Thyristor-controlled rectifier units are well suited to MMA and TIG welding, where the facility of accurate remote control is of value.

Transistorised units

The word 'transistor' is now well known thanks to the use in everyday life of transistorised radios, TV sets, and cassette players. Perhaps the most noticeable feature of these is their compactness. In welding power units, it is the way in which transistors operate that has opened up a totally new approach. Transistors can be used either as amplifiers or as switches.

When it is used as an amplifier, the transistor regulates the flow of current in response to a command signal. At first sight, it might be thought that this is similar to the thyristor unit, but the difference is very significant. In the case of the transistor, the current flows throughout the half cycle but its amplitude is reduced. This gives an output current which has less ripple than that from the thyristor unit.

A considerable amount of heat is generated in the transistor, and cooling is essential. There is also a limit to the amount of current that one transistor can carry and, for the current levels in welding, a large number of transistors are wired in parallel. Often there can be as many as 100 to 150 transistors in a pack. These power units are bulky and are very expensive, but they give very accurate control of the welding current and voltage.

The alternative is to use the transistor as a switch. In its simplest form, a bank of transistors can be used to switch resistors in or out of the circuit, altering the current (fig. 4.22). This is a useful technique at the low currents used for welding thin foil – say up to 10 A – but at high currents the resistors are expensive and need considerable cooling. One may successfully remove the resistors and use the transistor as a rapid switch. The switching frequency is about 5 kHz, and the mean current output depends on the ratio of 'Off to 'On' time. Heating in the transistors is less of a problem, as there is some cooling in each 'Off' period.

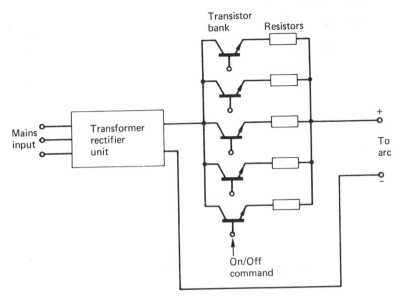

Fig. 4.22 *Switched transistor and resistor control of current*

Perhaps the greatest attraction of transistorised units is the ability to monitor the output to the arc and make adjustments via the command signal to the transistors (fig. 4.23). In this way the effects of mains-voltage variations can be offset. It is also possible to select either a drooping characteristic or a flat-characteristic output from the one unit. Last and by no means least, the output can be programmed – a facility which is of value in robotic applications (page 176).

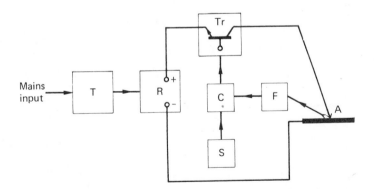

Fig. 4.23 *Elements of a transistorised power-supply unit to give either a drooping characteristic or a constant-potential output*
T – transformer; R – rectifier; Tr – transistor regulator; A – arc;
F – feedback voltage and/or current from arc; S – reference setting;
C – command unit (compares signals from F and S; amplifies error to give command signal for Tr)

Inverter systems

Although transistorised units overcome many of the drawbacks associated with conventional power sources, they represent a high capital investment which may be difficult to justify except in specialised applications or research projects. Two problems they fail to address are their weight and lack of portability. To a large extent this is because they still use an iron-cored transformer. A totally different approach, which enables the mains transformer to be dispensed with, is based on the use of an inverter.

In essence, an inverter is a high speed switching device. Supplied with d.c., it can be made to produce a series of pulses (fig.4.24), the duration and frequency of which are accurately controlled by a trigger mechanism which switches the inverter from non-conducting to conducting electricity and back again many times per second. Using a bank of four inverters, the wiring can be configured so that the output is a series of pulses which are alternatively positive and negative; in other words it changes d.c. into a.c.

Fig. 4.24 *Pulses from inverter*

Fig. 4.25 *Welding rectifier with inverter*

One critical feature of this conversion is the ability to set the frequency of the a.c. not at 50 Hz, as in a mains supply, but at far higher values ranging from 5 kHz to 25 kHz. The significance of this becomes apparent when we consider the design of the transformer. For applications with a.c. at 50 Hz, the core of the transformer is made from laminated sheets of iron. In contrast, at 5 kHz and above, the core can

Fig 4.26 *Portable inverter power unit*

be made from ferrite which is much lighter. In addition, at the higher frequencies the windings are made from smaller diameter wire and are more compact.

Applying these concepts to a power source for welding (fig. 4.25), the heavy mains transformer is dispensed with and replaced by a bank of rectifiers connected directly to the mains to give a 420 V (or 240 V) direct current. This is converted to high frequency a.c., fed to a lightweight transformer which reduces the voltage to that required for the arc and then rectified to give a d.c. output. A filter is included with the final rectifier to smooth out the ripple. The use of inverters in this way enables very compact power sources to be made. An inverter set for welding at currents up to 130 A is shown in fig. 4.26 – it weighs only 8 kg and can be carried with a strap over the shoulder.

Inverter units offer other advantages besides the saving of weight. By controlling the waveform produced by the inverter, compensation can be made for fluctuations in the mains supply voltage. Similarly, the arc voltage can be monitored and corrections made to the output to maintain a steady arc length. A logical next step from this is to build a memory in the controller and to programme it so that a pre-set arc voltage condition is readily established. Inverter systems have provided the key to the development of "one-knob" sets for MIG welding (page 100).

Perhaps the most significant aspect of these developments in the design of power sources for welding is the ability to make rapid corrections to the arc in response to information from monitoring devices. This not only leads to greater stability and consistency, it also means that the distinctions between output characteristics discussed earlier in relation to conventional units are no longer of concern with newer designs. A constant current or constant voltage mode of operation can be simulated by choosing a suitable method of controlling the electronic circuits.

5 Operating characteristics of manual metal arc (MMA) welding

Undoubtedly the most widely used arc-welding process is MMA. The plant is relatively cheap and simple, the welder has considerable freedom of movement, since the point of welding can be as much as 20 metres from the power-supply unit, and it is possible to weld a variety of metals with the same equipment simply by changing the electrode type.

5.1 Welding current

Either direct or alternating current can be used for MMA welding. To some extent, the choice is based on experience and individual preference, but a number of factors need to be considered:

a) Virtually all MMA electrodes work on d.c., but only certain flux compositions give stable operation with a.c.

b) Transformers are easier to maintain than the generators or rectifiers used for d.c. Also, a.c. units are more robust.

c) A d.c. arc may be deflected from the line of the axis of the electrode. It may be directed to the sides of the joint preparation or it may be bent forward or backwards in the direction of travel (fig. 5.1). This behaviour, called *Arc-blow*, is caused by the presence of magnetic fields around the arc, and these fields

Fig. 5.1 Arc-blow in MMA welding with direct current

can seriously disrupt effective operation. Arc blow only rarely occurs in a.c. welding, however, and it is associated with magnetic metals such as iron (steel) and nickel.

There are two main sources of magnetic fields in welding operations. The steel plate, pipe or section may become magnetised during manufacture or as a result of fabrication procedures which precede welding. In this case, the most satisfactory solution to arc blow is to demagnetise the metal either by heat treatment or by electrical techniques such as creating alternating magnetic fields in the component.

Perhaps the more common source of magnetic fields is the passage of the welding current through the parent metal. The direction in which the arc is moved by the magnetic forces depends on the location of the connection for the welding return cable. The solution usually lies in repositioning the connection. Some improvement can be obtained by wrapping the welding cable around the work.

d) Higher open-circuit voltages are required to maintain an a.c. arc.

The arc is extinguished each time the current goes through zero as the polarity is reversed, i.e. every one-hundredth of a second (fig. 5.2). If the weld pool is to remain molten, the arc must be instantaneously re-ignited. This requires that a voltage in excess of 80 V is applied to the electrode each time the current falls to zero. These high voltages can constitute a safety hazard (see appendix A) and d.c., with its lower OCV of about 60 V, is often preferred for this reason alone. Modified power supply units and new types of fluxes offer the prospect of using lower OCVs with a.c. welding.

Fig. 5.2 *Current and voltage waveforms in a.c. welding*

78

5.2 Functions of flux covering

The main reason for using a flux covering in MMA is to protect the molten metal from atmospheric contamination. At the same time, the flux fulfils a number of functions, all of which contribute to the success of the welding operation.

Weld-metal protection

The flux melts in the arc, along with the core wire. It covers the surface of the molten metal, excluding oxygen and nitrogen. When it has solidified, the flux forms a slag which continues to protect the weld bead until it has cooled to room temperature. While it is important to ensure that a good flux/slag covering is achieved, the flux composition must be chosen with a view to slag detachability, as the slag must be completely removed before the next run is laid. Ideally we would like a slag which lifts off the completed weld by itself. This is difficult to reconcile with the need for the slag to adhere to the weld metal during the cooling period, to avoid entry of air which would oxidise the surface. Slag detachability is also influenced by compounds included in the covering to achieve other objectives. The result is always a compromise, and the ease with which the slag can be removed varies from one electrode type to another.

Additional protection against atmospheric contamination is provided by including compounds which decompose in the heat of the arc to form gases. These replace air in the arc atmosphere, thus reducing the risk of oxygen and nitrogen absorption. They may be carbonates giving off carbon dioxide, or cellulose which produces an atmosphere of hydrogen and carbon monoxide.

Arc stabilisation

Although we often talk of the need to ensure that the arc is stable, it is very difficult to define what we mean by this. From an operational point of view, the easiest explanation is that we want the conditions to remain constant along the joint, unless the welder decides to make a change to accommodate some variation in fit-up etc. This implies that the top of the arc should always be at the centre of the core-wire cross-section, and the arc column should be in line with the axis of the electrode. In a groove, such as a single-V joint, the arc must not move from one side-wall to the other seeking the shortest path from the electrode to the work; it must stay firmly fixed in the direction dictated by the welder. At the same time, the end of the electrode must melt uniformly and the metal must be transferred to the weld pool without upsetting the stability of the arc.

Another aspect of stability is the ease with which the arc can be established at the start of a weld run or re-ignited at the beginning of each half cycle when using a.c. In both cases, the gas in the arc gap must be ionised rapidly and at the lowest possible voltage. Ionisation is facilitated by inclusion of titanium oxide, potassium silicate, and calcium carbonate in the flux. Silicates and oxides which have been added principally for other reasons can also act as arc stabilisers.

Control of surface profile

To understand the action of the flux in controlling the profile of a weld bead, we must first look at the role played by the surface tension of the weld pool. If the surface tension is high, the molten-metal surface becomes convex. On a flat plate, the liquid pulls away at the edges, and the angle of contact between it and the solid surface approaches 90°. This is another way of saying that the molten weld metal does not wet the solid parent metal. At the other end of the scale, if the surface tension is low, the contact angle is small and the molten metal has good wetting characteristics, so the surface of the pool is very flat.

Neither of these extremes is desirable in welding. A very high surface tension not only gives poor profile but also prevents the metal from flowing uniformly in the root of a 'V' groove. On the other hand, a very low surface tension makes it difficult to control the profile and to restrict the size of the pool. There is also a risk that the weld metal will flow over the joint faces before the arc has had time to melt them. Once again a compromise must be reached for satisfactory practical use.

The surface tension of the pool in arc welding is controlled by the oxygen level of the weld metal. This, in turn, is determined by the oxygen content in the flux. If the oxygen in the weld is low, this element will be transferred from the flux until a stable value has been established. The more oxygen there is in the flux, the higher will be the level in the weld. The effect this has on surface tension, and hence on surface profile, can be seen by examining fillet welds deposited in the horizontal – vertical position (Table 5.1). A low-oxygen weld has a high surface tension, and vice versa. In some cases, surface profile is not a prime consideration in formulating a flux, and a less-than-satisfactory profile may have to be accepted. This is the case with fluxes containing appreciable amounts of calcium fluoride and carbonate. These are used to give better mechanical properties which are associated with low oxygen contents. As we have seen, however, this means that the surface tension is high and we will inevitably have a convex profile if we use electrodes covered with these fluxes.

Table 5.1 Effect of flux oxygen content

Oxygen content	High	Medium	Low
Surface tension	Low	Medium	High
H–V fillet-weld profile			

Control of weld metal in position

The slag can be used as a mould which helps to keep the weld metal in place while welding in position. Three physical properties of the liquid slag must be kept in balance. First, it must have sufficient fluidity to flow freely from the root of the weld – to give good visibility and to avoid slag being trapped when the weld solidifies. The fluidity must not be too high, however, otherwise the flux runs off the face of the weld. The problem is eased if the surface tension is reasonably high, since this helps the slag to stay in place. Finally, the slag should solidify rapidly to form a solid barrier which restrains the tendency for the weld metal to run out of the joint.

Control of weld-metal composition

One of the outstanding advantages of MMA welding is that the electrode manufacturer can make adjustments to the composition of the weld metal by adding alloying elements to the flux covering. We have already noted that the oxygen content of the weld pool depends to a large extent on the amount of this element present in the flux. Similarly, manganese is added to the flux, in the form of ferromanganese, it will transfer into the weld. The actual amount which ends up in the weld depends partly on the concentration of manganese but also on the composition of the flux. For any given combination of flux and weld metal, alloying elements are distributed between the two in a more-or-less fixed proportion. The transfer can, of course, work both ways. If the flux or slag layer is low in manganese, this element transfers from the weld until a balance is established. Thus elements can be both added to and taken from the weld simply by altering the flux composition.

A good degree of control can be exercised over this transfer mechanism. During manufacture the thickness of the covering on the electrode can be maintained to within close tolerances. Hence the ratio of flux to core wire is reproducible and the amounts of alloying elements which need to be added to produce a particular weld-metal composition can be calculated by the electrode manufacturer. There are three main aspects of weld-metal composition control which should be considered: alloying, deoxidation, and contamination.

Alloying

If a core wire is used which has the same composition as that desired in the finished weld, it is necessary only to ensure that there is no loss into the flux – there is no need to add alloying elements. This is common practice with non-ferrous metals and some stainless steels.

With electrodes for use on low-carbon, carbon, carbon-manganese, and low-alloy steels, alloyed core wires tend to be expensive and it is preferable to carry out the alloying in the weld pool. In other words, the core wire is a low-carbon steel, and alloy elements – e.g. manganese, chromium, and molybdenum – are added to the flux. Thus the electrode manufacturer has a considerable degree of flexibility and enables a large range of electrodes to be produced with the same core wire. This is of particular advantage with alloy steels where small quantities of special-composition electrodes may be needed.

Deoxidation

If a molten weld pool in steel contains large amounts of oxygen, bubbles of carbon monoxide are formed which are trapped in the solidifying metal to form porosity. The gas arises from a reaction between the dissolved oxygen and carbon:

$$FeO + C \longrightarrow Fe + CO$$

The carbon monoxide is not soluble in steel and is released as a bubble. Quite apart from the formation of porosity, the carbon loss is undesirable as it has an influence on the strength of the weld. The reaction must be suppressed, and in MMA welding this is achieved by adding a deoxidant to the flux.

A deoxidant is an element for which oxygen has a strong affinity. When a deoxidant is added to a weld pool, the oxygen combines with it in preference to reacting with carbon. An oxide is formed which floats to the surface of the pool and mixes with the slag. In MMA welding of steel, the most commonly used deoxidant is silicon, which is added to the flux in the form of ferro-silicon. The deoxidising reaction in the weld pool is

$$2FeO + Si \longrightarrow 2Fe + SiO_2$$

Other deoxidants are used with different metals; for example, when welding copper it is usual to add phosphorous or zinc to remove the oxygen. These are not necessarily introduced via the flux and may be included in the core wire.

Contamination

So far we have discussed the use of additions to the flux to achieve desirable improvements in the weld. We must also recognise that the flux can introduce harmful elements.

In Chapter 7 we will be discussing the role of hydrogen in the formation of cracks in a welded joint. Hydrogen comes from the covering of the electrode, which normally contains both absorbed and chemically combined moisture. We can dry the electrode to reduce the level of absorbed moisture, but the amount of combined water present depends on the chemical compounds in the flux. Various ways have been used to measure the quantity of hydrogen present. The most useful technique for comparing electrodes is to deposit a weld run under controlled conditions and collect the hydrogen which diffuses out over a specified period of time (fig. 5.2). The result is expressed as the millilitres of hydrogen

Fig. 5.2 *Simple apparatus for collecting diffusible hydrogen*

in each 100 grams of weld metal; this is abbreviated to ml/100 g. Data obtained from this test can be used in the design of weld procedures to avoid cracking (see Chapter 7). The lowest hydrogen contents in MMA welding are obtained with electrodes containing large amounts of calcium compounds. These are discussed in more detail later, under the heading of Basic coverings (page 84).

Other contaminants introduced via the flux covering are usually present by accident. Grease and oil can be collected during storage. Damp sulphurous fumes may be absorbed if electrodes are left in the vicinity of pickling vats. All of these are transferred with the flux via the arc to the weld pool, where they are absorbed. Careful storage and handling of electrodes are therefore of the utmost importance if high-quality welds are to be deposited.

5.3 Functions of individual constituents of a flux

It will be appreciated that a large number of chemical compounds are used in formulating a flux. Table 5.2 lists the compounds used in MMA electrodes for the welding of steels and indicates the roles which each flux constituent plays.

Table 5.2 MMA electrode constituents and their functions

Constituent	Primary function	Secondary function
Iron oxide	Slag former	Arc stabiliser
Titanium oxide	Slag former	Arc stabiliser
Magnesium oxide	Fluxing agent	———
Calcium fluoride	Slag former	Fluxing agent
Potassium silicate	Arc stabiliser	Binder
Other silicates	Slag formers and binders	Fluxing agents
Calcium carbonates	Gas former	Arc stabiliser
Other carbonates	Gas former	———
Cellulose	Gas former	———
Ferro-manganese	Alloying	Deoxidiser
Ferro-chrome	Alloying	———
Ferro-silicon	Deoxidiser	———

Note Binders are used to give the flux covering mechanical strength and to help it adhere to the core wire. Fluxing agents are used to adjust surface tension and wetting characteristics.

5.4 Operating characteristics of electrodes for steels

There are four main groups of electrodes used in MMA welding of steels. They are distinguished by the major constituents of the flux covering which determine their operating characteristics.

Acid coverings

Acid coverings are composed mainly of oxides and silicates and have a high oxygen content. They give smooth weld profiles with a tendency to concavity.

The slag has a porous or honeycomb structure (inflated) and is easy to detach. Although they have good ductility, the welds tend to be low in strength, and for this reason acid electrodes are not widely used.

Cellulosic coverings

Cellulosic coverings have large quantities of organic material containing cellulose. Flour and wood pulp are common constituents. The organic compounds decompose in the arc to generate hydrogen which replaces air in the arc column. The presence of hydrogen increases the voltage across the arc and makes it more penetrating. For a given current, the depth of penetration with a cellulosic electrode is about 70% greater than with other types. Because most of the flux is decomposed, the resultant slag layer is thin and in some instances need not be removed before depositing the next weld run. The surface profile is poor and, while the mechanical properties are good, the hydrogen content is very high, which restricts the use of this type of electrode on high-strength steels.

Rutile coverings

Rutile coverings are based on titanium oxide. This compound has good slag forming characteristics and produces a stable easy-to-use arc. Rutile electrodes are widely used and fulfil a general-purpose role in the fabricating industry. The deposits have a medium oxygen content (page 80); hence surface profiles are acceptable, and slag detachability is good. By varying the additions of fluxing agents, the viscosity and surface tension can be adjusted to give electrodes which are suitable either for the flat position only or for use in all positions. Mechanical properties are adequate for most structural steels, but it is not easy to achieve high tensile strengths. The flux can be dried, but it will always contain a reasonable proportion of combined water which is needed to keep the covering intact. If this moisture is driven off, the binding of the flux will suffer. The hydrogen contents of the weld metal are high, being of the order of 25 to 30 ml/100 g. This is above the level normally considered desirable for high-strength steels.

Basic coverings

Basic coverings mainly contain calcium compounds such as calcium fluoride and calcium carbonate. The term 'basic' refers to the chemical behaviour of the flux. It does not mean that the electrodes are simple or easy to use.

They are sometimes called 'lime-coated' or low hydrogen electrodes and are used principally for the welding of high-strength steels. The flux can be dried by baking at approximately 480°C, and, if the electrodes are stored at 150°C until used, the hydrogen content of the weld metal can be reduced to between 10 and 15 ml/100 g. At this level the risk of cracking in higher-strength steels is minimised. Electrodes which can be treated in this way are called 'hydrogen-controlled'. The mechanical properties are very good, and it is possible to produce weld deposits which match almost all commercially used carbon-manganese and low-alloy steels. The oxygen content of the slag is low, however, and surface profiles of weld deposits are convex (page 80) The slag is difficult to detach, and more fumes are given off than with other types of electrode.

Iron-powder additions

Iron-powder additions are sometimes made to the flux covering to increase the electrode efficiency. This is defined as the mass of metal deposited as a percentage of the mass of core wire melted:

$$\text{electrode efficiency \%} = \frac{\text{mass of metal deposited}}{\text{mass of core wire melted}} \times 100\%$$

In general the efficiency of MMA electrodes is between 75 and 90%. Some of the metal is lost as fume and oxides in the arc, but the largest source of loss is spatter, i.e. droplets of molten metal ejected from the arc.

Electrode efficiencies rise above 100% if iron powder is added to the flux covering. Thus, in a deposit made with an electrode said to have an efficiency of 160%, slightly more than 60% of the weld metal will have come from the flux, the remainder being melted core wire.

In addition to giving high deposition rates, iron-powder added to the coating also tends to increase the oxygen content of the slag, producing welds with smooth flat surfaces. The slag detaches readily, and the electrode is easy to use in the flat or horizontal–vertical positions.

5.5 Current ranges for MMA electrodes

The core wires of electrodes for MMA welding are generally 450 mm long and have a diameter within the range 2.5 to 6.3 mm. As the electrical connection is made at the top of the electrode, the full welding current flows along the core wire (fig. 5.3). This core wire has electrical resistance, and heat is generated by the passage of the current. If the temperature rises too high, there is a danger that the

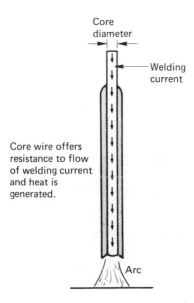

Fig. 5.3 *Current flow in an MMA electrode*

electrode will melt prematurely. Before this happens, however, the flux can be damaged. Moisture evaporated from the compounds in the flux causes it to flake off, leaving lengths of the core wire uncovered. At the same time, oxidation of some of the alloying elements can affect the composition of the resultant weld. With these points in mind, the electrode manufacturer stipulates a maximum current for each electrode diameter. There is also a lower limit of current, below which the arc becomes unstable. Table 5.3 gives typical ranges for MMA electrodes; the actual values depend on the type and composition of the flux and are specified by manufacturers for their own products.

Table 5.3 *Current ranges for MMA electrodes*

Core-wire diameter (mm)	Current (amperes)	
	Minimum	Maximum
2.5	50	90
3.2	65	130
4.0	110	185
5.0	150	250
6.0	200	315
6.3	220	350

5.6 Choosing and specifying an electrode

There is no single answer to the question 'which is the best electrode?' Each application must be evaluated by examining a number of aspects; for example

a) what is the composition of the metal to be melted?
b) is there a risk of weld metal cracking?
c) what mechanical properties are required?
d) is a.c. or d.c. plant available?
e) what is the position of welding?
f) what is the thickness of the parent metal?
g) what is the type of joint?
h) are there any limitations on heat input?

Having answered these and other questions, an electrode type is selected which gives the optimum performance at an economic price. It will then be found that the type of electrode chosen is marketed by a number of electrode manufacturers. Each uses a brand name for the electrode, making it difficult to compare one product with another. The task of comparison is made easier, however, by use of an electrode-specification system. Most industrialised countries have produced their own system – some are based on the international standard ISO 2560; others are unique to a particular country. Mostly the index number allocated to an electrode indicates the properties of the weld deposit, the type of covering, and the operating characteristics (i.e. type of current, a.c. or d.c., and best positions of welding). This can be illustrated by two examples:

Operating characteristics of manual metal arc (MMA) welding

a) United Kingdom: BS 639 (based on ISO 2560)

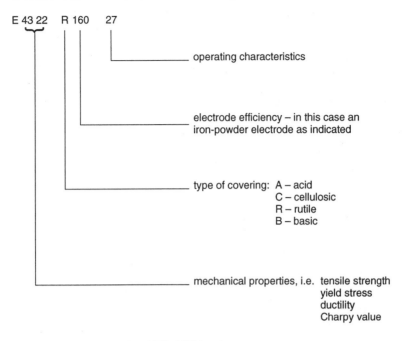

E 43 22   R 160   27

— operating characteristics

— electrode efficiency – in this case an
iron-powder electrode as indicated

— type of covering: A – acid
C – cellulosic
R – rutile
B – basic

— mechanical properties, i.e. tensile strength
yield stress
ductility
Charpy value

b) United States of America: AWS-ASTM code

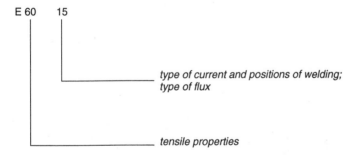

E 60   15

— type of current and positions of welding;
type of flux

— tensile properties

6 Operating characteristics of gas-shielded welding

6.1 Tungsten arc gas-shielded (TIG) welding

Perhaps the outstanding feature of TIG welding is that the tungsten electrode is non-consumable. As there is no metal transferred across it, the arc is used solely to melt the parent metal. If additional metal is required to build up the weld surface, fill a groove, or form a fillet in a T joint, it can be added to the weld pool separately in the form of filler wire. The degree of control which this arrangement gives the welder makes the TIG process particularly suitable for the welding of sheet material. Welds can be deposited in all positions, but the level of skill required is very high. Control of the weld pool is also important in the deposition of root runs in pipe. TIG welding is often used for this purpose in the fabrication of pipework for high-pressure steam lines and for chemical plant (fig. 6.1).

Argon supplied at a pressure of 1.1 bar to protect the underside of the weld.

Removable blanks confine backing argon to the area of the weld.

$R = 5$ mm

2 mm

15°

If extra metal is required it can be added manually in the form of filler wire at the leading edge of the pool.

or

Consumable inserts can be preplaced in the joint — these are melted into the weld pool.

Fig. 6.1 *TIG welding of root runs in pipe*

Welding current

Direct current with the electrode connected to the negative pole of the power-supply unit is used for all metals except aluminium. This arrangement gives a stable arc. About 60% of the total heat is developed at the surface of the work, thus the larger proportion of the heat is available to melt the parent metal. Some 10% is lost by radiation and only 30% is released at the electrode; hence it is relatively easy to keep the electrode below its melting point (3370°C). The end of the electrode is tapered to encourage the arc to operate from the tip. This improves the arc stability and helps the welder to keep the arc length constant.

Aluminium and its alloys present a major problem. When the parent metal is melted, the oxide skin remains intact as its melting point is appreciably higher than that of aluminium. As there is no flux in TIG welding, the oxide skin must be dispersed by the action of the arc. This happens only when the electron flow is from the weld pool, i.e. when it is negative and the electrode is positive. Under these conditions, oxide-free welds can be produced in aluminium. The main drawback is overheating of the electrode, which occurs because the biggest proportion of the heat is being developed in it, i.e. it is the reverse of the situation we discussed above with the electrode negative. This poses so many difficulties that electrode-positive d.c. is rarely used in practical applications.

The solution lies in the use of an a.c. arc. During the half cycles when the electrode is positive, the oxide is dispersed from the molten aluminium. The electrode cools during the negative half cycles while heat is built up in the weld pool. In this way a uniform distribution of heat is achieved and the electrode can be prevented from melting to any great extent – usually the tip becomes a hemisphere of molten tungsten (fig. 6.2).

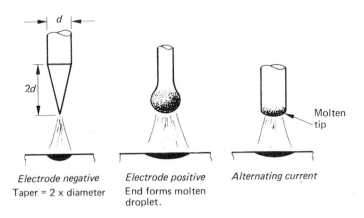

Electrode negative   *Electrode positive*   *Alternating current*
Taper = 2 x diameter   End forms molten droplet.

Fig. 6.2 *Electrode configuration in TIG welding*

Electrodes

Although a pure tungsten electrode can give satisfactory results, it is often preferable to use an electrode which contains additions of either thoria (thorium oxide) or zirconia (zirconium oxide) to improve arc striking and stability.

Thoriated electrodes are used for d.c. welding, while zirconiated electrodes are preferred for a.c. The diameter of the electrode is chosen to suit the current (Table 6.1).

Table 6.1 *Typical maximum currents for TIG electrodes*

Electrode diameter (mm)	Maximum current (amperes)	
	Thoriated (d.c.)	Zirconiated (a.c.)
1.2	70	40
1.6	145	55
2.4	240	90
3.2	380	150
4.0	440	210

Starting the arc

By contrast with MMA welding, the arc is not started by touching the electrode on to the plate. In the TIG system this would contaminate the tungsten with the parent metal and lower its melting point. A method of arc initiation must be incorporated which allows the welding current to flow across the gap between the electrode and the work. This requires ionisation of the gas in the gap, which can be achieved by injecting a high frequency (h.f.) discharge or a high-voltage spark. An h.f. unit is essentially a tuned-oscillator circuit consisting of a transformer, a capacitor, and a spark gap (fig. 6.3). The tuned circuit operates at a frequency between 300 kHz and 3 MHz. The spark starter uses a high-voltage coil of the type found in car ignition circuits, with the arc gap replacing the sparking plug.

Fig. 6.3 *High-frequency arc-starter unit for TIG welding*

Once the gas in the gap has been ionised by either the h.f. or the spark discharge, the full welding current flows and the arc is established. With d.c. welding, the arc starter can be switched off at this point. On the other hand, in a.c. welding the arc is extinguished each time the current passes through zero going from negative to positive and vice versa. The arc must be re-ignited one hundred times each second. This means that the arc starter must operate continuously throughout the welding operation.

Rectification in a.c. arcs

Ideally the current flow in the welding circuit must be the same during the negative and positive half cycles. If there is a difference, one of two problems can arise. A reduction in the current during the negative half cycle leads to overheating of the electrode. On the other hand, if the positive half cycle is restricted, the oxide on the pool is not adequately dispersed. In both cases, the result is similar to that which would be produced if a direct current was superimposed on the alternating current. For this reason the imbalance is referred to as rectification (fig. 6.4).

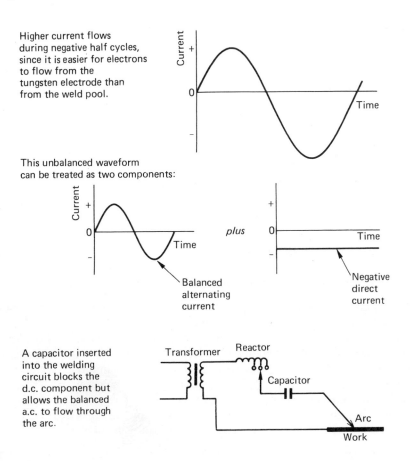

Higher current flows during negative half cycles, since it is easier for electrons to flow from the tungsten electrode than from the weld pool.

This unbalanced waveform can be treated as two components:

plus

Balanced alternating current

Negative direct current

A capacitor inserted into the welding circuit blocks the d.c. component but allows the balanced a.c. to flow through the arc.

Fig. 6.4 *Inherent rectification in TIG welding with alternating current*

In practical welding operations the most common form of rectification is that resulting from the difference between the arc voltages during the negative and positive half cycles. A higher voltage is needed when the electrode is positive, with the result that a lower current flows. This is known as 'inherent' rectification and is offset in conventional power sources by including in the welding circuit a large capacitor which allows the a.c. to pass but blocks the d.c. component (fig. 6.4). In units fitted with solid state systems, inherent rectification can be offset by electronic means.

Shielding gas

In TIG welding, pure argon is normally used as a shielding gas. It works well with both d.c. and a.c. and is compatible with all the weldable metals. The nozzles used are designed to match the flow characteristics of argon so that there is no turbulence (fig. 6.5). This gives protection to the top of the weld pool, but the underside in a root run remains exposed to air and can oxidise rapidly. Where this is a problem, a secondary supply of argon is provided on the reverse side of the joint. In pipe welding, the bore of the pipe is 'purged' i.e. it is filled with an inert gas such as argon (fig. 6.1).

Fig. 6.5 *Gas shielding in TIG welding*

Helium is an inert gas which provides an alternative to argon. Arc voltages are higher when a tungsten arc is shielded with helium. This is not to be confused with raising the arc voltage by increasing the arc length. For a given arc length, the voltage across the gap is higher when it is filled with helium than in the case of an argon shield. The result is increased heat input, giving improved welding on metals having a high thermal conductivity. The depth of penetration for a given current is also greater, resulting in higher travel speeds.

Helium is significantly more expensive than argon, and it is usual to use a mixture of helium and argon. Two mixtures are readily available:

75% helium-25% argon for aluminium and copper plate and sections;
30% helium-70% argon for aluminium and copper sheet.

Although the mixed gases are more expensive than pure argon, the cost can be offset by savings in operation.

One application where the use of pure helium can be justified is the welding of thick aluminium plate. The fierceness of the helium shielded arc coupled with its penetrating characteristics breaks up the oxide film and enables d.c. to be used. Welding speeds when using the TIG process on stainless steel or nickel alloys can be increased by adding 1% to 5% hydrogen to the argon shielding gas. This also gives deeper penetration and a less oxidised surface on the finished weld.Gas flow rates depend on the design of the gas nozzle and the type of joint; they are normally about 5 to 7 litres per minute (l/min) for welding sheet but may be increased to 10 l/min for thicker material.

Low current operation

TIG welding is mainly used on sheet material less than 4 mm thick; above this, travel speeds become very slow. The lower limit of thickness is deter mined by arc stability. At currents less than 10 A the arc wanders from point to point on the surface of the weld pool and drifts from the tip of the electrode. Stable operation is possible if the current is supplied in pulses. The pulse frequency varies from 10 per second to 1 per second (fig. 6.6). The current during the pulses is at a level where the arc is stable. The heat input depends on the mean current, which is controlled by the pulse height, pulse frequency, and pulse duration. The weld consists of a series of overlapping circular spots, each of which is melted by one pulse and freezes before the next pulse. Pulsed TIG welding can be used not only on thin sheet but also in root runs, as it gives better control of penetration than conventional continuous current techniques.

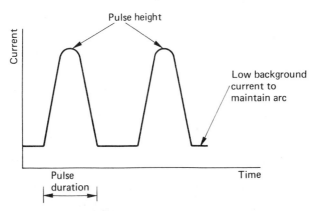

Fig. 6.6 *Pulsed current supply for TIG welding*

6.2 Metal arc gas-shielded (MIG) welding

In many ways MIG welding is a direct competitor with the MMA process. It offers faster welding in similar applications, but uses more expensive equipment and consumables. The quality of the welds is comparable, and the choice between the two is often based solely on the relative operating costs. MIG welding has opened up new areas of work in the sheet-metal industry for which MMA was not suitable. It plays an important role in the production of car bodies, where freedom from frequent electrode changing and from the need to remove flux are important production factors. In thick-plate applications, the high rate at which electrode metal is deposited can help reduce production costs.

Electrode wires

Although the absence of flux is an advantage in the welding operation, it does mean that all the alloying elements and deoxidisers must be present in the electrode wires. Electrodes range in diameter from 0.8 to 1.6 mm and are specially produced for MIG welding, as the surfaces must be free of contamination.

Steel wires contain manganese and other elements such as chromium, molybdenum, and nickel to give strength or corrosion-resistance to the weld metal. The amounts added are usually slightly larger than the corresponding parent-metal contents to allow for alloy loss in the arc. Silicon is used to deoxidise the weld pool as in MMA welding, but must now be included in the wire itself. Aluminium is added to some electrodes to act as a deoxidant. It reacts with oxygen at a faster rate than silicon but it can leave a detrimental effect on the ductility of the weld metal where the sulphur content of the parent metal is high enough to form aluminium sulphide in the weld. Silicon contents range from 0.4 to 0.9% and aluminium, if added, is usually about 0.15%. The wires are copper-coated to provide better electrical pick-up in the contact tube and to reduce the risk of corrosion of the wire in storage.

Electrode wires for stainless steel, weldable aluminium alloys, and other non-ferrous metals usually match the plate in composition. They are not copper-coated.

Flux-cored electrodes

In contrast to electrodes for MMA welding, the alloying elements in solid wires for MIG welding can only be added during the manufacture of the ingots from which the wires are drawn. This does not pose a major problem in aluminium and other non-ferrous metals. With steel, however, it leads to high cost; steelmaking operates most efficiently on large tonnages, whereas manufacturers of wires for MIG welding handle relatively small amounts. The compositions of the wires are different to the general run of steel used for plate, sheet etc. and the steelmaker has to produce small melts which increase the cost. Finally, some steels used for MIG wires are difficult to draw down to small diameters.

Flux-cored electrodes provide a possible answer in some applications. The electrodes are, in effect, long thin tubes into which fluxes and alloying elements are packed (fig.6.7). During welding these are melted into the weld pool in a similar manner to the MMA process. The electrode can be a simple tube or there may be strips folded into the centre to increase the amount of metal in the cross-section. The outside diameter of the electrode ranges from 0.9mm to 3.2mm.

In addition to providing alloying elements and deoxidisers, the flux can also produce a slag which assists in shaping the profile of fillet welds in the horizontal-vertical position. Fillet welds with leg lengths of the order of 10mm can be difficult to control in bare wire MIG welding.

Flux-cored electrodes are suitable for use on a range of steels, including some grades of chromium-molybdenum steels, stainless steels and abrasion-resistant steels. Pure CO_2 gas is widely used for welding with flux-cored electrodes. For positional welding better results may be obtained by using an argon-20% CO_2 gas

94

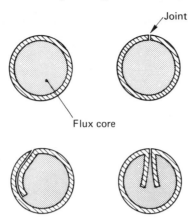

Fig. 6.7 *Cross-sections of typical flux-cored electrodes*

together with a suitable small diameter electrode. Self-shielded electrodes are available and are particularly useful for site work. The core of the electrode contains gas generating compounds which, in conjunction with the flux, protect the arc and eliminate the need for additional gas shielding.

Shielding gases

The shielding gas in the MIG system is varied to suit the parent metal (Table 6.2). To some extent this is done to reduce costs, but more often it is to achieve better arc operation. Once again, the nozzles must be designed to give a non-turbulent flow of shielding gas over the weld area. The flow rates are significantly higher than with TIG – 12 to 20 litres/min compared with 4 to 6 litres/min.

Table 6.2 *Shielding gases for MIG/MAG welding*

Metal	Shielding gas
Aluminium and alloys	Pure argon
Nickel and alloys	Argon + helium
	Pure argon
Copper	Argon + helium
Stainless steel	Argon + 3% oxygen
Low-carbon steel and carbon-manganese steel	Argon + helium + CO_2
	Carbon dioxide
	or
	argon + 5 to 20% CO_2
	or
	argon + 5% $°2$
Steels with 1 to 2% chromium	Argon + 5 to 20% CO_2
	or
	argon + 5% 0_2
Steels with more than 2% chromium	Argon + 5% 0_2

This is a suitable point at which to consider the name of this process. In the early stages of its development, MIG welding was called semi-automatic because, unlike MMA and TIG welding, both the electrode feed and the arc length were controlled by the welding equipment. At the same time a significant feature of the process was the inert gas shield. In the USA, where the process was developed, it was called Gas Metal Arc Welding (GMAW) to distinguish it from Gas Tungsten Arc Welding (GTAW). At the time the only gas being used for shielding was argon, and in the UK the title Metal Inert Gas (MIG) was adopted. The list of gases in Table 6.2 shows that active gases are also used for shielding. This raises no problems with the US term GMAW, but it creates an anomaly in the use of MIG. The alternative Metal Active Gas (MAG) has been introduced to overcome this, but it still creates a difficulty where reference is being made to the

Heater to vaporise liquid carbon dioxide

Release valve

Combined regulator and flow gauge

Regulator

Gas supplied to torch at 2 bar pressure

Vapour pressure forces liquid up the tube to the heater

Liquid carbon dioxide

Syphon tube

Argon + oxygen
or
argon + carbon dioxide
(steel cylinder containing
7 m³ of gas at
175 bar pressure)

Carbon dioxide
(stored as liquid in
steel syphon-type
cylinders)

Fig. 6.8 *Cylinders used for argon and carbon-dioxide gases in MIG welding*

process in general terms rather than being specifically related to either inert or active gas. In this context it is helpful to combine both abbreviations, i.e. MIG/ MAG, if we are not dealing with a specific type of gas.

Pure argon is used for aluminium and its alloys, nickel and its alloys, and many non-ferrous metals. Helium added to the argon increases the heat input, and the mixture is used to advantage when welding metals with a high thermal conductivity, such as copper. With ferrous metals such as stainless steels, an arc operating in pure argon tends to be unstable. The low oxygen content of the arc atmosphere also causes the surface of the pool to become convex and to pull away from the edges, leaving the weld depressed or undercut (fig. 7.1). An addition of 1 to 5% oxygen to the shielding gas overcomes this problem. Subsequently the oxygen combines with silicon in the electrode wire to give a glass-like slag which floats to the surface of the molten pool.

Carbon dioxide is also used to shield the arc when welding low-carbon steel and carbon-manganese steels (CO_2 welding). Although this gas is chemically active, in the arc it breaks down to a mixture of oxygen and carbon monoxide. Extra silicon added to the electrode removes the oxygen, leaving the weld pool blanketed by carbon monoxide which does not react with molten steel. A mixture of argon and carbon dioxide (5% to 20%) is also used. The choice between argon + 5% oxygen, argon + carbon dioxide, and pure carbon dioxide as a shielding gas for the welding of steels is not simple and is based on a comparison of arc behaviour, spatter levels (i.e. the amount of metal ejected from the arc as molten droplets), weld profile and penetration, and cost. The cost aspect is affected by supply, i.e. whether you are buying individual cylinders of gas (fig. 6.8), or large containers of liquefied gas, the quantity purchased, and whether the gas is supplied pre-mixed or is mixed at the user's factory.

The use of carbon dioxide, by itself or mixed with argon, is often restricted to the welding of steels without chromium. Carbon dioxide reacts with chromium to form a carbide, thus reducing the effective content of this alloying element in the steel. Shielding gases are usually sold under trade names. As with MMA electrodes, it can be difficult to compare one manufacturer's gas with another's, as the trade name does not necessarily indicate the composition of the gas. In Europe a proposed classification has been devised based on an alphanumeric code. The gases are divided into groups which reflect their chemical behaviour – gases can be inert or they can create an oxidising or reducing atmosphere.

Table 6.3 *Classification of gases (prEN 439)*

Group code	Characteristic
R	reducing gas mixtures
I	inert gas mixtures
M	oxidising mixtures containing oxygen, carbon dioxide or both
C	highly oxidising mixtures
F	unreactive gas; this classification allows for gases which are strictly reactive in chemical terms but behave as an inert gas in the arc, for example, nitrogen which can be used to weld some grades of copper.

The classification letter is followed by a number which identifies the composition, for example:

Argon + 5% carbon dioxide	M1
Argon + 10% hydrogen	R1
Argon + 10% hydrogen + 40% helium	R1(2)

Gas flow rates are generally higher than those used for TIG welding, ranging from 10 l/min at currents used for dip transfer to 15 l/min at 300 A to 400 A (spray transfer).

Welding current

For commercial applications, MIG/MAG welding uses d.c. with the electrode positive. Two current ranges are normally used. The higher range covers from 300 A upwards and is confined to joints in plates in the flat and horizontal vertical positions. The heat input is too high for positional work, and the penetration is too great for manual welding of sheet. These applications are better suited to the low-current range (50 to 180 A), where it is possible to produce a small weld pool

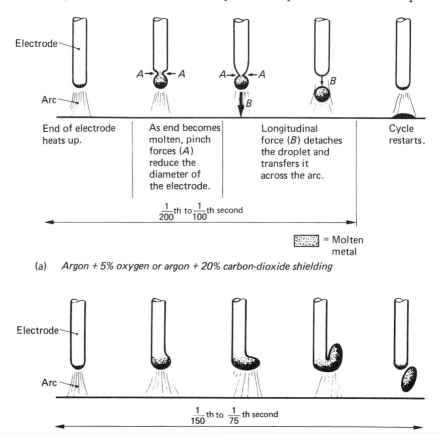

Fig. 6.9 *Spray transfer in MIG/MAG welding*

which freezes quickly and has less penetration. Aluminium provides an exception to this in so far as positional welding is concerned. The high current range extends from 220 A and can be used for welding vertically and overhead. The pool is retained in position by an oxide skin which forms over the surface. At the same time, heat is conducted away rapidly from the fusion zone, since aluminium has a high thermal conductivity. The arc is very stable at high currents. Droplets are detached regularly from the end of the electrode and are transferred to the pool by electromagnetic forces acting in the arc along the line of the axis of the electrode (fig. 6.9). This is known as *spray transfer*. In argon shielding, the droplets are detached completely from the electrode before they reach the weld surface. On the other hand, in carbon-dioxide shielding the droplet becomes very large before being detached and may actually touch the weld pool, creating a short circuit. It is the rapid rise in current during this short circuit which leads to smaller droplets being ejected from the arc area as spatter.

Arc heats weld pool.
Electrode tip is moving towards surface of pool.

Direction of welding

Arc length gets shorter since current is not high enough to produce rapid melting of electrode.

Tip of electrode touches the weld pool. Power-supply output is short-circuited and the current rises.

The rise in current is controlled so that the end of the electrode is resistance-heated.

Heated region

End of electrode melts and flows into the weld pool.

The arc is re-established and the sequence is repeated.

Time for complete sequence $= \frac{1}{200}$th to $\frac{1}{50}$th second

Fig. 6.10 *Dip transfer in MIG/MAG welding*

If we wish to weld in position or to join sheet metal, we must reduce the heat input. However, at low currents, i.e. less than 200 A, the forces in the arc are too small to detach and transfer the droplets. The arc is unstable and, unless the mechanism of transfer is altered, the system cannot be used for welding. The problem can be solved in two ways. In the first, known as *dip transfer*, a short arc is established by lowering the voltage to about 21 V. At currents below 200 A, the end of the electrode melts slowly. As the electrode is being fed into the arc at a steady rate, the arc gap is shortened until the tip touches the weld pool (fig. 6.10). The output from the power supply is now being short-circuited, and a very high current flows through the electrode. If this were allowed to continue, the wire would melt and be ejected as spatter. The inclusion of an inductance in the welding circuit controls the rate at which the current rises, so that the end of the electrode is melted uniformly and flows into the weld pool, re-establishing an arc gap. This cycle is repeated 50 to 200 times each second.

The alternative technique is known as *pulsed transfer* (fig. 6.11). The arc is operated with a normal arc length and a low background current of the order of 50 to 100 A. At this current, heat is maintained in the weld pool and some electrode

Fig. 6.11 *Pulsed transfer in MIG welding*

heating takes place. Pulses of high current (in excess of 300 A) are superimposed on to the low background current. During each pulse, the arc behaves as if it were operating in the spray range and a single droplet is transferred before the current is returned to a low level. This gives spray-type transfer at a low heat input, because the latter is proportional to the mean current. The most commonly used pulse frequency is 50 Hz.

Both dip and pulsed transfer enable MIG/MAG welding to be operated at low current. This means that positional welding and the joining of sheet material down to 0.8 mm thickness are possible. Dip transfer is by far the most commonly used technique but it is applicable only to metals which have a relatively high electrical resistance, i.e. steels and some nickel alloys, since the electrode is melted by resistance heating during the short circuiting period.

Improvements in the stability and control of both dip transfer and pulsed-arc systems are possible with solid-state power supply units (see page 72). Dip transfer in particular is critically dependent on the maintenance of a steady voltage supply. Variations in voltage output from the power source can lead to areas of lack of fusion. With a solid state unit fluctuations in mains supply voltage can be monitored and the output voltage adjusted to compensate. The ability to make automatic corrections whilst welding (page 76) also offers the possibility of using predetermined relationships between voltage, current, wire diameter and wire feed rate. The logical extension is to provide a built-in set of relationships from which one is selected to suit a given application i.e. 'one-knob' control.

In pulsed MIG it is also possible to determine the optimum pulse current level, duration and frequency for a given combination of material, shielding gas, and arc voltage and current level. A 'library' of these relationships can be stored in a memory ready to be selected to suit a particular application. Having established the parameters at the beginning of a weld run, monitoring devices maintain the relationship or select different combinations to ensure consistency in the operation of the arc. The term used to describe this technique is 'synergic' (working together). A library of conditions for a given combination of metal and shielding gas can be assembled, based on trials, and stored in a memory ready to be selected for a given application. The task of selection can be reduced to the operation of a single control – 'one-knob' welding sets.

7 Weldability

7.1 Response of metals to welding

Most metals are weldable in that it is possible to create a molten pool between two pieces of metal which solidifies to give a weld. However, there are notable exceptions; for example, with zinc-based castings and some aluminium based die-casting alloys, melting cannot be confined to the joint area and the weld pool becomes too large to control. Some metals are easier to weld than others. Steel is readily weldable; in many ways it is easier to weld than aluminium or copper. The film on the surface of aluminium hinders fusion of the parent metal (page 89). Copper is generally regarded as not easily welded because the high thermal conductivity makes it difficult to raise the parent metal to its melting point; it is often necessary to preheat the joint to between 300°C and 400°C.

Although aluminium and copper are weldable, it does not necessarily follow that they have good weldability. The latter concept involves far more than just creating a weld between two pieces of metal.

7.2 Defining weldability

The term *weldability* is not easily defined. It takes into account the extent to which the metal is weldable in the terms discussed above, but it also incorporates a consideration of quality. Detailed aspects of quality will be considered in Chapter 8. For the present it is sufficient to recognise that it is unusual for welds to be completely sound i.e. totally free of defects or discontinuities (fig 7.1). More often than not they contain small amounts of defects such as porosity (small holes or gas pores), slag or oxide inclusions, and lack of fusion (areas at the junction between the weld metal and the parent metal where bonding has not taken place). These are faults that relate to the technique used to make the weld and to a large extent are a measure of the skill of the welder, who must aim, of course, to keep faults to a minimum. It may be more difficult to control defects with some metals. For example, with the type of steel used for sheets which are to be pressed into shapes for car bodies (Table 7.1 – rimming steel) porosity may be difficult to avoid. Similarly, oxide inclusions are a persistent problem in the welding of aluminium, resulting from inefficient removal of the surface film. These "process" defects are not, however, a characteristic of the metal and are not part of weldability. Cracks, on the other hand, are mostly related to the composition of the material being welded and the risk of their formation is an important element in the concept of weldability. Indeed, in many specifications weldability is taken as being synonymous with the tendency to cracking.

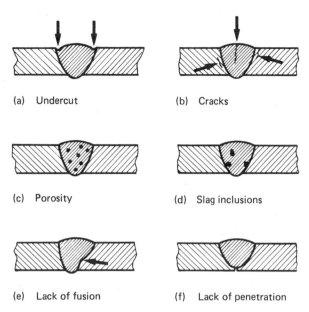

Fig. 7.1 *Cross-sections of welds containing typical defects*

In assessing weldability we are also concerned, however, with properties. These may be mechanical, as in tensile strength and Charpy value (Chapter 8). They can also be electrical, or chemical, as in oxidation or corrosion resistance; wear resistance may be another factor to be taken into account. When properties are the principal consideration, statements about weldability may appear to be contradictory. An example of this is to be seen in the welding of aluminium alloys containing magnesium and silicon (table 7.3 – 6082). Risk of cracking in the weld is minimised by using a filler wire rich in silicon. The resultant weld metal has a tensile strength which is significantly lower than that of the parent metal, and so the alloy could be said to have poor weldability. On the other hand, in applications where strength is not a consideration, the corrosion resistance of the joint may be sufficient to make it acceptable.

So, a useful working definition of weldability is the ease with which a welded joint can be made free of cracks and with properties which match a given specification. The question which now arises is what governs the properties of and the risk of cracking in a welded joint.

7.3 Properties of a welded joint

In Chapter 2 (page 20) we noted the distinction in grain structure between weld metal and parent metal. The former is composed of dendritic grains which have grown from the molten pool on solidification whereas the parent metal is made up of more or less uniform grains which reflect its manufacturing history. The mechanical properties of a metal depend critically on the grain structure, so we must expect that they differ from the parent metal to the weld metal. The grain structure is also modified by heating, so we need to examine the effect of the welding operation on the parent metal alongside the weld.

Table 7.1 *Classification of steel*

Steel is an alloy of iron and carbon with additions of other metallic elements chosen to give specific properties.

● Carbon and Carbon Manganese Steels

Sometimes called mild steel – deprecated term – or structural steel
Carbon (up to 0.25%) plus manganese; may have small amounts of silicon.
Used for structural work, shipbuilding, pressure vessels and pipework.

● Medium Carbon

Carbon about 0.4%.
Mainly used for engineering components such as shafting.

●Low Alloy

Chromium, nickel, molybdenum and vanadium are added in small amounts to achieve specific properties. Only a few are readily weldable, for example:
   Chromium-molybdenum steels for steam piping;
   Nickel (9%) steels for low temperature applications.

● Corrosion Resistant (Stainless) Steels

Contain chromium (13% to 25%).
Nickel (8% to 20%) added to certain grades.

● Manufacturing method

Steels can also be classified by manufacturing method. During the steel making process, the molten metal absorbs oxygen. When the ingot has been cast and starts to solidify, the oxygen reacts with carbon in the steel. This has two consequences. The carbon content is reduced and gas bubbles are formed. The amount of oxygen in the molten metal is controlled by adding a deoxidant immediately before the molten steel is poured into the ingot moulds. Manganese, silicon and aluminium are the most common deoxidants.

● Low carbon steel (sometimes called rimming steel)

Deoxidants are not added. The gas bubbles are trapped in the ingot and the carbon content is reduced to low levels. Rolling the ingot to sheet closes the gas holes and welds them together. The resulting sheet has very good ductility. This type of steel is widely used in the manufacture of cars and domestic items such as washing machines.

● Killed steel

An excess of deoxidant is added to the molten steel so that the carbon content is not reduced significantly. Other alloying elements - for example chromium, nickel and vanadium - are retained in the steel. Typical examples are higher strength carbon-manganese steels for use in pressure vessels and low alloy steels.

● Semi-killed steel

Deoxidant additions are restricted to lower levels. Some gassing takes place in the ingot but carbon contents are retained. Semi-killed steels are cheaper to produce than killed steels but usually have lower strength. Structural steels are usually of the semi-killed variety.

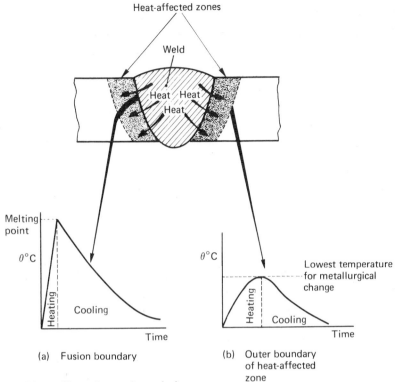

Fig. 7.2 *Heat-affected-zone boundaries*

The heat affected zone

As the molten weld starts to cool, only a small amount of heat escapes through the surface of the pool. The majority of heat flows through the parent metal on each side of the joint. As a result, the parent metal is subjected to a heating and cooling cycle (fig.7.2). The two critical components of the cycle are the highest temperature reached and the subsequent rate of cooling. The maximum or peak temperature depends on the distance from the weld. Parent metal near the fusion boundary is heated to just below the melting point. Moving away from the fusion boundary, the peak temperature falls progressively until at some point the parent metal ceases to be heated. Before this is reached, the temperature will have fallen below the lowest value at which a relevant metallurgical change takes place. The precise value of this temperature depends on the composition of the metal and the type of change being considered. The region of the parent metal which has undergone a metallurgical change as a result of the thermal cycle is called the *heat affected zone* (h.a.z.).

The effects of the heating and cooling cycle in the h.a.z. can be illustrated by three types of metallurgical reaction: softening of work-hardened material, softening of precipitation hardened alloys and hardening of steels. The significance of these metallurgical changes in terms of the acceptability of a welded joint are discussed in Chapter 8; here the aim is to examine the factors which influence the metallurgical behaviour of the metal in the h.a.z..

Work-hardened metals

Work-hardened metals are characterised by increased hardness and tensile strength, resulting from deformation (working) at room temperature. Typical examples of work-hardened materials used in welded fabrications are pure aluminium and its alloys containing manganese and magnesium (page 112), and steel sheet with thicknesses of 0.7 mm or 0.9 mm (22swg or 20swg) which is used for the manufacture of domestic appliances such as refrigerators. Work-hardening of the sheet results from the rolling process which is used to reduce the thickness of the metal. In the earlier stages of rolling from an ingot or from thick plate, the metal is kept at a high temperature to achieve large reductions in thickness. The rolling is finished with the metal at room temperature (cold working) to give greater control of accuracy and a better surface finish. Cold working also has the effect of increasing hardness and tensile strength. Often a designer will specify work-hardened material to gain advantage from the higher strength. Work-hardening can also occur in the fabrication shop as a result of pressing, bending or forming at room temperature.

The effects of cold working – depending on the composition of the metal – are removed if the metal is heated to a critical temperature and softening occurs. This is known as recrystallisation. During welding, metal which is raised to this temperature is recrystallised, and as a result, the h.a.z. contains softened metal (fig.7.3) which has a lower tensile strength than the unaffected parent metal. The properties which have been lost in this way cannot be recovered.

Precipitation-hardened alloys

Precipitation-hardened alloys derive strength from a controlled heat treatment which is often called age-hardening. The hardening is associated with the formation of small groupings (precipitates) of the alloying elements which are distributed uniformly throughout the alloy. Examples of metals which can be hardened by precipitation treatment are aluminium alloys (page112), some grades of stainless steel, and nickel alloys such as Nimonic. The treatment

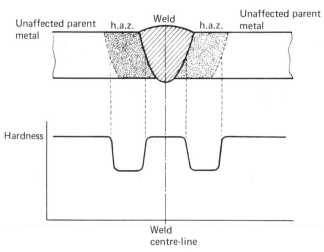

Fig. 7.3 Hardness variations in a welded joint in work-hardened material

consists of heating the sheet or section to a high temperature, which depends on the composition of the alloy, and cooling it quickly. It is then reheated to an intermediate temperature and held there for a prolonged period during which the hardness of the metal increases as the precipitates are formed. After a while the precipitates grow to a critical size and the hardness reaches a maximum value at which point the metal is cooled rapidly. The heating cycles experienced in the h.a.z. of a welded joint disperse the precipitates, returning the metal to the softened condition. In contrast to h.a.z. softening of work-hardened metals, the hardness and strength of a precipitation-hardened alloy can be recovered by heat-treating the finished weld. One particular type of alloy is worthy of mention in this context. Aluminium alloys containing typically 4.5% zinc and 1% magnesium (7020) recover their hardness after welding without the need for heat-treatment. If they are left at room temperatures the hardness increases over a period of about 14 days. This behaviour is known as *natural ageing*.

Hardening of steels

Thus far we have been concerned with the peak temperature reached in the h.a.z. With steel the critical factor is the cooling rate (fig.7.2). In heat-treatment practice steels are hardened by heating to a temperature around 1,000°C and then cooled rapidly by plunging the metal into water or oil (quenching). During the heat-treatment, the metallurgical structure of the steel is changed from a ductile to a hard, brittle form. The metallurgist uses terms such as martensitic and bainitic to describe this hardened type of structure. The amount of hardening depends on the speed of cooling - fast rates are needed to achieve hardening - and the composition of the steel. The cooling rates experienced in the h.a.z. of a welded joint are often high enough to give quenching effects similar to those experienced in heat-treatment practice to produce hardening.

The fact that the h.a.z. in a welded joint in steel is hard is not in itself a problem. The main concern is that it creates conditions that encourage the formation of cracks in the h.a.z. (fig 7.4). There is a high risk of cracking if three

Fig. 7.4 H.A.Z. cracks in a high-yield-strength steel

conditions exist. The first is that the hardness is above a critical level. The second is the presence of tensile stress, and the third is that there is hydrogen present in the weld pool. It is important to note that, even if these conditions are met, it is not certain that cracks will be formed. There are a number of factors such as the type of hardened structure, the joint profile and the actual level of hydrogen which interact in a complex way. We talk, therefore, of the *risk* of cracking.

The level of hardness reached in the h.a.z. may be as high as 450VPN; it is determined by the cooling rate (fig.7.5). Hardness is measured by forcing an indentor into the surface of the steel and comparing the size of the impression with an arbitrary standard. In the most commonly used system the indentor is a diamond and the hardness is quoted in Vicker's Pyramid Numbers (VPN). A carbon-manganese steel typically used in fabrication would have a hardness of 190 to 200VPN as delivered from the supplier. This corresponds to the so-called *normalised* condition which means that the steel has been heat-treated to remove any hardening that might have occurred during rolling, and to produce a uniform grain size which is essential for optimum properties.

When the hardness is 350VPN or higher there is a high risk of cracking; in other words the structure is *crack sensitive*. The cooling rate (fig.7.6) at which this level of hardness is produced is related to the composition of the parent metal. Steels used in welded fabrications contain a number of alloying elements in addition to carbon (table 7.2). Some of these appear to play little part in the formation of h.a.z. cracks, but others, such as manganese and chromium, increase the ease with which a hardened structure can be formed. In assessing their significance it is convenient to use an index known as the *Carbon Equivalent* (CE). In this, the amount of each alloying element present is factored according to the contribution it makes to hardening. Thus, manganese is allocated a factor of one sixth as it has been estimated that a manganese content of 0.6% produces the same effect on hardening as 0.1% carbon. There are many

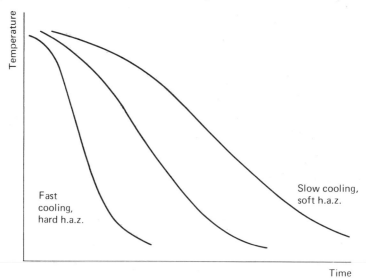

Fig. 7.5 *Effect of cooling rate on h.a.z.*

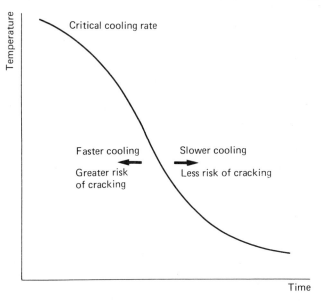

Fig. 7.6 *Critical cooling rate in h.a.z. of weld in steel*

Table 7.2 *Alloying elements in steel*

- **Carbon**
  main alloying element which determines the strength and structure of the steel.
- **Manganese**
  increases strength and hardness - steels for structural work, bridges and pressure vessels contain between 0.8% and 1.6% manganese.
- **Silicon**
  used to deoxidise steel in manufacture, and is usually present only as a residual element
- **Chromium**
  increases strength, especially at temperatures up to 300°C, and improves corrosion resistance. It also increases hardening in the h.a.z.
- **Nickel**
  added with other alloying elements to increase the strength and improve the impact properties at low temperatures - a 9% nickel steel is used for storage tanks containing liquids at 162° below freezing.
- **Molybdenum**
  improves creep strength - between 0.5% and 1.5% molybdenum together with 1.5% to 5% chromium.
- **Sulphur**
  is an impurity, but amounts up to 1% can be added to improve machinability.
- **Phosphorus**
  is an impurity.

variations of the formula used to calculate the CE. They all have the same format but differ in the factors used. The formula used in British Standards is:

$$CE = C\% + \frac{Mn\%}{6} + \frac{(Cr + Mo + V)\%}{\%} + \frac{(Ni + Cu)\%}{15}$$

where: C = carbon, Mn = manganese, Cr = chromium Mo = molybdenum
V = vanadium Ni = nickel and Cu = copper
In structural steels, CE values range from 0.35% to 0.5%.

The interaction of cooling rate and carbon equivalent is illustrated in fig.7.7. At low levels of CE, fast cooling rates can be tolerated before there is a risk of cracking; except in thick sections, h.a.z. cracking is rarely experienced with CE values below 0.39%. At high levels of CE, say 0.48%, there is a very high risk of cracking even with slow cooling rates.

However, it does not necessarily follow that cracks will be formed when the hardened h.a.z. is subjected to tensile stresses caused by heating and cooling of the joint. The deciding factor is the presence of hydrogen. This gas is absorbed into the molten weld pool from a variety of sources: moisture in the flux covering of an electrode or in the shielding gas, grease on the joint faces and so on. Hydrogen can flow (diffuse) readily through hot steel and passes from the weld pool into the h.a.z. If the concentration of hydrogen is high, there is a major risk of cracking. Conversely, if there is no hydrogen present, the possibility of a crack forming is negligible. Gas shielded processes such as MAG and TIG are inherently low in hydrogen with levels ranging from 5 ml/100 g to 10 ml/100 g; the method of measuring hydrogen levels is described on page 82. Levels of hydrogen in weld metal deposited by the MMA process range from 10 to 30 ml/100 g depending on the type of flux covering (pages 83 to 85).

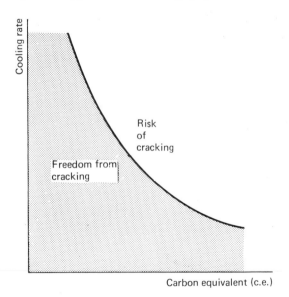

Fig. 7.7 *Interaction of cooling rate and carbon equivalent*

110

The cooling rate in the h.a.z. is primarily determined by the heat input and the thickness of the metal in the joint. High heat inputs result in slower cooling rates in the h.a.z. In thick sections the cooling rate is faster than in thin. For example, a T-joint in 30mm thickness plate in a particular grade of steel may have a high risk of cracking, whilst a butt weld in 6mm plate with the same composition may be deposited without any cracks in the h.a.z. The cooling rate is also influenced by the temperature of the parent metal before and during the welding operation. Preheating a plate to, say, 150°C and keeping it at that temperature while the weld is being deposited, slows down the rate of cooling through the temperature range within which a hardened structure is formed - typically 300°C to 200°C. Preheat also helps to reduce the risk of cracking by allowing any hydrogen in the heat affected zone to flow into the parent metal where hardening has not taken place.

The principal factors governing the risk of h.a.z. cracking in carbon steels are:

● composition (carbon equivalent)
● cooling rate (heat input and joint type and thickness)
● hydrogen content
● preheat (temperature of parent metal during welding).

The interdependence of these factors is complex. A simplified diagrammatic account of the relationships is given in fig.7.8. In practice, the choice of welding

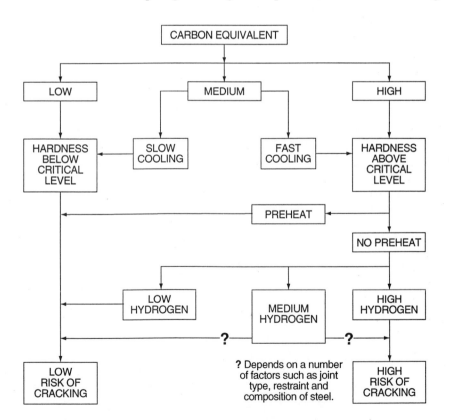

Fig. 7.8 *Principal factors involved in formation of h.a.z. cracks in steel*

111

parameters to establish a situation in which the risk of cracking is minimised is not easy. Guidance is provided by documents such as British Standard BS 5153.

Weld-metal

In Chapter 2 we noted that the weld-metal had a dendritic or columnar structure. This is typical of cast metal and in general is associated with lower ductility than the equivalent wrought material (parent metal) which has a uniform, equi-axed structure (pages 20and 24). With most metals used in fabrication, the weld metal has lower strength than the parent metal. Steel is an exception in that yield stress is higher in the weld itself. The design aspects of strength of welded joints is discussed in Chapter 8.

The pattern of solidification described on page 20 assumed that the metal was pure, i.e. it did not contain alloying elements. The majority of metals welded are alloys - a fact which has important implications in relation to cracking in the weld metal. As the solidification proceeds, the molten metal becomes enriched with not just alloying elements but any impurities which were in the parent metal or on the surfaces of the joint. This effectively lowers the melting point, with the result that a liquid film remains between the dendrites (fig.7.9). As the shrinkage stresses build up (page 144) because there is no cohesion at the grain boundaries, the boundaries are pulled apart and a crack is formed. Fortunately, this does not occur in all metals. In general carbon steels are not prone to this problem. The exception is when the steel contains an appreciable amount of sulphur. This can occasionally come from contamination of the joint faces. Usually, however, the problem is met in a particular grade of steel to which the steelmaker has added sulphur to improve the machining properties: free-cutting steel. There is little that can be done to minimise this problem, and sulphur bearing free-cutting steels are not normally recommended for arc welding.

With other materials, however, such as some aluminium alloys (table 7.3) or austenitic stainless steel (table 7.4), the formation of cracks in the weld metal can be prevented by using filler metal of a different composition to the parent metal. The aim is to prevent the formation of low melting point films so that the weld has sufficient strength to withstand the shrinkage stresses. As an example, heat-treatable aluminium alloys, particularly those containing copper, suffer from weld metal cracking. In TIG and MIG welding this is controlled by using a filler metal which is high in silicon. Two grades are in common use: 4043 (5% silicon) and 4047 (10% silicon). With austenitic stainless steels the filler wire or electrode composition is much closer to that of the parent metal. The choice of a suitable

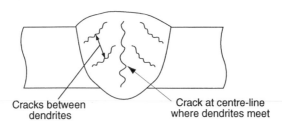

Cracks between dendrites

Crack at centre-line where dendrites meet

Fig. 7.9 *Weld metal cracking*

Table 7.3 *Aluminium alloys*

Aluminium can be alloyed with a number of metals to give a wide range of properties. The alloys can be divided in to two main groups: non-heat-treatable and heat-treatable. The type of alloy is identified in British Standards by a four figure code. The first digit indicates the alloy group. The ISO (International Standards Organisation) code uses a combination of letters and numbers. The letters are the chemical symbol for the alloying element, e.g. Al Mg2 designates an aluminium (Al) alloy with 2% magnesium (Mg) addition.

● **Non-heat-treatable Alloys**
There are two main groups of non-heat-treatable alloys:
Aluminium-manganese (3XXX) (Al Mn)
Aluminium-magnesium (5XXX) (Al Mg)
They are supplied in either the softened condition or with different levels of work-hardening, designated quarter, half, three-quarters or full hard. Alloys in both groups are readily weldable by TIG or MIG and are widely used in fabrication. Commonly used grades are 3103 (Al Mn1), 5251 (Al Mg2) and 5083 (Al Mg4.5 Mn).

● **Heat-treatable Alloys**
The main groups of heat-treatable alloys are:
Aluminium-copper (2XXX) (Al Cu)
Aluminium-magnesium-silicon (6XXX) (Al Mg Si)
Aluminium-zinc-magnesium (7XXX) (Al Zn Mg)

● **Aluminium-copper Alloys (2XXX)**
(sometimes called Dural which is a trade name) suffer from severe weld metal cracking and are generally considered not to be weldable.

● **Aluminium-magnesium-silicon alloys**
e.g 6060 (Al Mg Si) and 6082 (Al Si Mg Mn), are used for pipe and extruded sections. They are weldable but require an aluminium-silicon filler (4043 or 4047) to avoid weld metal cracking; the resultant weld metal has low strength.

● **Aluminium-zinc-magnesium alloys**
are readily welded. The most common alloy, 7020 (Al Zn4.5 Mg1), is used for structural work and military applications such as bridges. The properties of the joint can be recovered after welding by heat-treatment. Alternatively, the structure can be left in the as-welded condition in which case the hardness increases with time to nearly match the parent metal value (natural ageing).

Table 7.4 *Stainless steels*

The term stainless steel is used for a number of alloy steels which show a high resistance to corrosion. There are three main groups based on the metallurgical structure of the steel.

● **Martensitic Stainless Steels** contain 13% Chromium.

High carbon martensitic stainless steels are hard and are used for cutlery and similar applications where a cutting edge is required. Welded joints are prone to severe h.a.z. and weld metal cracking.

Low carbon varieties do not harden to the same extent as the high carbon martensitic steels, and the low carbon varieties can be formed and welded. Used widely in the petrochemical industry for pipes and liners for vessels.

● **Ferritic Stainless Steels** contain 17% chromium and are mainly used for pipework in petrochemical plant, especially where sea water corrosion is a problem. They are weldable, but precise control of fit-up and heat input are essential.

● **Austenitic Stainless Steels** are the most widely used variety of stainless steel. They are often called 'eighteen-eights'; strictly refers to only one alloy in the range (Grade 321 - see below). Chromium contents range from 18% to 25%. Nickel (8% to 20%) improves ductility, and they are readily formed or shaped. Other alloying elements are added to confer resistance to specific corrosive environments. Austenitic steels cannot be hardened by heat treatment.

The grades of austenitic stainless steel which are usually found in welded fabrications are:

Grade	Chromium%	Nickel%	Others
304L	18.5	10.0	
309L	22.0	12.0	
316L	17.0	12.0	Molybdenum
321	18.0	9.0	Titanium
347	19.0	10.0	Niobium

● **Duplex Steels** are so called because their structure is a mixture of ferrite and austenite. They contain chromium (18 to 27%), nickel (4 to 6.5%), molybdenum (1.5 to 3%) and nitrogen (0.1 to 0.35%) and are used in highly corrosive environments containing chlorides. They are readily weldable.

Risk of centre-line cracking
if $\frac{D}{W} > 1.5$

Fig. 7.10 *Centre-line cracking in single weld pass*

filler to control cracking in stainless steels is more complex than is the choice with aluminium alloys as the former also involves consideration of the type of application.

A particular type of problem which is associated more with the profile of the weld is found in steels. When the weld run is deep and narrow, the direction of growth of the dendrites tends to be towards the centreline rather than upwards to the weld surface (fig 7.10). The result is that the metal at the top of the weld solidifies leaving insufficient molten steel to fill the shrinkage cavity which is developed at the vertical centre line, so forming a cavity or crack. Deep, narrow welds are characteristic of high current consumable electrode welding systems such as MAG and SAW and result from the use of an arc voltage which is too low for the current being used. The solution is either to increase the arc voltage or, if this is not practicable, to limit the depth of penetration by reducing the current.

8 Weld quality and strength

8.1 Criteria for weld acceptability

Welds are not of themselves either good or bad, although sometimes examples may be produced which are so full of holes and inclusions that it is difficult to imagine an application for which they could be used. When we refer to a weld as being 'bad', we really mean that it is unacceptable for the purpose we have in mind and we must recognise that a standard or quality which one person might regard as reasonable may be too high or too low for other fields of work. Take, for example, a butt joint in a nuclear reactor. Once the reactor is working it might be possible, although undesirable, to carry out repairs to a weld which develops a fault. Of far greater importance is the fact that the failure of a critical joint could have disastrous consequences and it would be quite unjustifiable to risk putting into service a weld which contains any significant defects. In a storage tank, on the other hand, one of the main considerations would be the corrosion-resistance of the weld, and defects which would be a cause for immediate rejection in the case of the nuclear reactor could well be considered harmless in the tank, since high strength is not always a prime requirement for service.

While it is relatively easy to make general comparisons of the type given above, it is much more difficult to write a precise definition for the acceptable quality of a weld. Usually we attempt to relate this definition to the presence of defects such as porosity, undercut, lack of bonding between weld metal and plate, cracks, and pieces of slag trapped in the weld. It is unfortunate the term 'defect' has been used in this context, since it implies that the very presence of one or more of these makes the weld defective, but, as we have seen, there are cases where they can be tolerated. In fact the basis of a specification for weld quality is frequently an assessment of the number and size of the defects which can be present in a weld before it is considered to be defective and therefore rejectable. This is not to say that we should aim to produce welds containing defects – on the contrary, the welder should always use procedures and techniques which avoid the formation of the faults noted – but, because reality is not perfect, we know that defects will occur, and our standard of acceptance is simply stating the point at which corrective treatment must be applied.

The designer of a component or structure may have to take into consideration a number of criteria, such as:

mechanical properties	resistance to corrosion
resistance to oxidation	resistance to abrasion
leak tightness	vacuum tightness
stiffness.	

The acceptance standard for the welds must reflect the significance of each defect in relation to those design criteria which are of importance. Mostly, however, we are concerned with how welds behave under various types of tensile loading.

8.2 Strength of butt-welded joints

The strength calculations made in the design of a fabricated structure are based on the properties of the metal being used. Ideally the weld should have properties at least equal to those of the parent metal. In practice, the weld has a cast structure (page 110) and with most of the metals used in welded constructions it has a lower tensile strength than the parent metal. The principal exception is structural steel where the yield stress of the weld is normally higher than that of the parent metal. There must be a clear understanding of the effect of welding on the strength of the joint, so that the designer can allow for this in the calculations. It may also be that with certain types of loading, for example fatigue, the presence of a welded joint automatically reduces the permissible maximum stress. We can start with a butt joint in which there is bonding between the weld metal and parent plate at all points throughout the thickness. It is useful to make the proviso that the excess metal and penetration bead should be ignored, since we cannot control their dimensions precisely (fig. 8.1). The load-bearing thickness of the butt weld – the throat – is now equal to the thickness of the parent metal. We have two cases to consider, i.e. when the weld-metal yield stress is firstly greater and secondly less than the yield stress for the parent metal.

We can see the importance of the weld-metal yield stress by looking at the behaviour of a test-piece during a tensile test in which the test-piece contains a weld. The weld is transverse to the applied stress and has been machined flush with the surface of the test-piece. As the weld metal and parent material are essentially similar, they have the same modulus of elasticity and stretch uniformly as the load is applied in the elastic range. If the weld metal has the higher yield stress, yielding occurs first in the parent metal at some point away from the weld, accompanied by necking. The true stress in this region increases more rapidly than the stress in the weld metal, since the cross-section is being reduced by the necking, and fracture takes place in the plate. This is the basis for the often expressed view that a weld repair is stronger than the original part, but this applies only to tensile loading in those circumstances in which a weld metal with a higher yield stress can be used as in the case of carbon-manganese steels.

With steel, we are usually able to select filler metal having suitable mechanical properties, i.e. we can overmatch the parent metal. With many non-ferrous

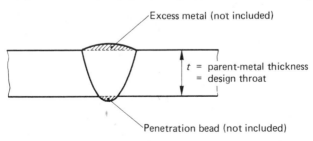

Fig. 8.1 *Design throat in a butt joint*

metals and alloys this is often not the case. In the first place, these materials tend to be weaker if they have a cast structure – in the weld for example. It may also be necessary to use a filler-metal composition which gives a lower strength in order to avoid or minimise cracking in the weld metal. Undermatching is therefore commonplace with non-ferrous metals, and, when a tensile load is applied, yielding first takes place in the weld metal. Fracture occurs in the weld, and the tensile strength of the joint is below that of the parent metal. The extent of this difference can be expressed in terms of joint efficiency:

$$\text{joint efficiency} = \frac{\text{tensile strength of joint}}{\text{tensile strength of parent metal}} \times 100\%$$

Summarising, in the case where yielding in the tensile test occurs away from the weld, as with overmatching, the allowable design stress is that of the parent metal. On the other hand, with joint efficiencies of less than 100 percent, the margin between the parent-metal design stress and the stress at which yielding occurs is reduced, and some adjustment to working-stress levels must be made.

8.3 Strength of fillet-welded joints

The fillet-welded joint requires a very different approach because, unlike the butt weld, the size of the fillet is controlled by the welder and the throat dimension can be increased at will. In a T joint, the throat is taken as the shortest distance from the surface of the weld to the intersection of the weld and the joint at the root (fig 8.2). This includes the excess metal and, as with the butt weld, it is customary for design purposes to neglect the contribution this part of the weld makes to strength. We must, therefore, redefine the design throat as the shortest distance from the root to the line drawn between the toes of the weld. It is assumed that

Fig 8.2 *Throat thickness and leg length in fillet welds*

fracture under load will occur across the throat. In practice, weld-metal failures in fillet-welded joints under test often tend to follow a line closer to the fusion boundary, but the error involved in using the design throat as defined above is relatively small. The point is somewhat academic, since our aim is to design a joint which will not fail under the loads experienced in service.

The load-carrying area of the fillet weld is the design-throat thickness multiplied by the length of the weld. In fillet welds which transmit loads, the throat area is in shear and the maximum allowable shear stress will be determined by the composition of the weld deposit, which in turn depends on the type of electrode used and the material being welded. As an example of this, the information for fillet welds in Table 8.1 is taken from British Standard BS 2573: part 1, which is concerned with crane jibs made from carbon-manganese steels.

Table 8.1 Strength of fillet welds

Plate properties (N/mm^2)		Max. allowable shear stress on fillet welds (N/mm^2)	
Tensile strength	Min. yield stress	Electrode E43	Electrode E51
430–510	245	118	126
500–620	355	118	144

The calculation of the size of fillet for a given application is based on the concept that the throat area should be large enough to transmit the maximum tensile stress allowed for the parent metal. The length of the weld is usually dictated by the dimensions of the component or structure, and the calculation aims to specify the minimum throat thickness. This dimension is of little direct help to the welder, since it is difficult to estimate the throat thickness while depositing the weld, especially if it contains more than one run. It is more normal to quote the leg length in instructions given to the welder.

8.4 Effect of heat affected zone properties

The way in which the thermal cycle in the h.a.z affects mechanical properties was discussed in the previous chapter (p104).

With structural steels, the parent metal in the h.a.z. either remains more or less unaffected or it is hardened. As we have seen, the principal concern with a hardened structure is the risk of cracking which would severely curtail the joint's performance in service. If the h.a.z. remains free of cracks, the hardening does not enter into the assessment of the tensile strength of the joint.

The situation is very different with materials which have been hardened either by mechanical working, such as cold rolling, or by heat treatment. With these materials, the strength in the h.a.z. is reduced to that of the unhardened or annealed condition. For example, a 5083 Aluminium-magnesium alloy, which is widely used for cryogenic plant and ship superstructures, has a tensile strength of about 340 N/mm^2 in the work-hardened condition. After welding, the strength in the h.a.z. would be as low as 275 N/mm^2, representing a significant loss of

119

properties in the welded joint that could not be recovered. Similarly, a 6082 Aluminium-magnesium-silicon alloy, used in pipes, structural sections and extrusions for windows, has a tensile strength of 295 N/mm², whereas after welding, the h.a.z. strength is about 210 N/mm². It would be possible to heat the joint and recover the strength in the h.a.z., but practical constraints may make such an operation difficult. The use of a 7020 alloy may provide an answer to this problem in some applications. A further consideration is the strength of the related weld metal. Many precipitation hardened alloys are prone to weld metal cracking (page 112). To avoid this in a 6082 alloy, the filler wire usually contains between 5% and 11% silicon, resulting in a weld strength of about 180 N/mm².

8.5 Influence of weld defects

In the discussion of the strength of a weld (page 116) it was assumed that the weld metal was completely free of defects. This is a rarely achieved ideal. Most welds made under factory or site conditions contain at least one or two small defects. Six typical defects were illustrated in figure 7.1 (page 103).

The first thing to notice is that in a weld cross-section these defects provide discontinuities in the path of a transverse tensile stress, and it would be tempting to suggest that their significance is in the reduction of the effective load-bearing area. In fact, in many cases a reduction of up to 28% in cross-sectional area could be tolerated before yielding would take place in the weld rather than the plate, and this reduction would need to be along the complete length of the weld. This might occur in the case of incomplete penetration, but it is most unlikely that undercut, porosity, cracks, slag inclusions, or lack of fusion could produce such a significant loss of cross section. We must therefore look to another reason why we do not accept defects.

Stress concentrations

Consider a plate subjected to a tensile load (fig. 8.3). The stress σ is uniformly distributed across the plate and is given by:

$$\sigma = \frac{\text{load } (F)}{\text{area } (A)} = \frac{F}{\text{width } (w) \times \text{thickness } (t)}$$

If a hole of radius a is drilled at the centre of the plate, the width of the plate is reduced by $2a$. The load is therefore being carried by a smaller cross-section and the stress rises to:

$$\sigma_2 = \frac{F}{(w - 2a) \times t}$$

This is the mean stress level; the actual stress distribution is not uniform but is high at the edges of the hole, falling to the original level al some distance away.

The discontinuity in the path of the stress flow created by the hole has caused a concentration of stress. The severity of this stress-concentration effect depends on the shape of the discontinuity: with a round smooth hole the concentration factor is low, being of the order of 3, while a sharp notch can give stress concentrations as high as 10. The real significance of defects is their ability to

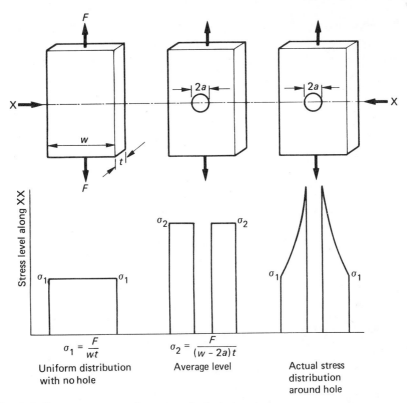

Fig. 8.3 *Stress concentration around a hole in a bar*

create stress concentrations. This means that we can be more tolerant of rounded discontinuities such as gas pores and slag inclusions than of defects such as cracks. Table 8.2 illustrates the acceptance criteria that might be specified for a pressure vessel. Reference must be made to the appropriate standard for specific allowances, e.g. BS 5500 .

Table 8.2 *Typical allowance for defects*

Defect type	Acceptance criteria
Porosity-isolated	Diameter can not be greater than 6 mm, depending on plate thickness.
Porosity-distributed	2% to 3 % of the area of the weld metal as seen on a radiograph.
Slag inclusions	Length can be numerically equal to plate thickness.
Incomplete root penetration	Not permitted.
Lack of fusion	Not permitted.
Cracks (any type)	Not permitted.

Notch ductility

In making a full assessment of the significance of a defect, reference must be made to the material in which it exists. A crack transverse to the direction of applied tensile stress (fig. 8.4) creates a high concentration at its ends.

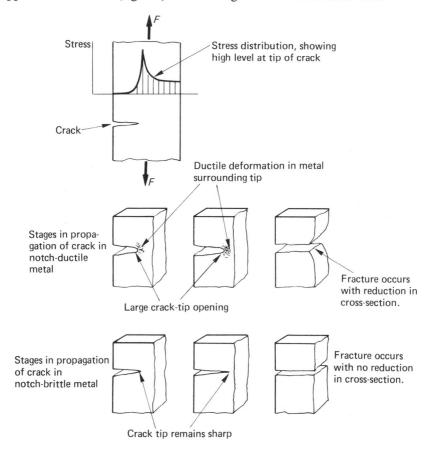

Fig. 8.4 *Comparison of notch-ductile and notch-brittle behaviour*

In ductile material, yielding takes place in the material surrounding the tip of the crack, and the width of the crack increases at the same time as the crack extends in length. The continued behaviour of the crack depends on a number of factors, and in certain circumstances the crack may not propagate. If it does proceed across the complete width of the component, the final fracture shows evidence of appreciable plastic deformation. On the other hand, in some materials there is very little opening of the crack, which then propagates very rapidly across the specimen, giving a fracture in which there has been virtually no reduction of area, i.e. a brittle failure. Such a material is said to be notch brittle, as opposed to the notch-ductile case quoted before; alternatively, it can be described as having low notch ductility.

The problem of notch-brittle behaviour is a very real consideration in the design of structures fabricated in carbon or carbon-manganese steels, since these materials can exhibit both notch-ductile and notch-brittle behaviour, depending on temperature. A series of Charpy tests (fig. 8.5) performed at different temperatures ranging from, say, –60 to +60°C on specimens cut from steel plate gives the sort of results shown graphically in fig. 8.6. Although there is usually considerable scatter, a clear trend can be observed from a high level of energy absorbed during fracture with a marked fall to very low energy values.

The temperature at which this transition takes place is a characteristic of the composition of the steel, its grain size, and the treatments it has received. Below the transition temperature a planar defect such as a crack or lack of fusion behaves in a notch brittle manner and a fast-running brittle fracture could occur. It is because of risks of this type that standards such as BS 5500 are severe in their treatment of planar defects, while other standards concerned with less stringent situations are more prepared to accept, say, lack of fusion. In general, cracks of significant size are never tolerated.

Establishing standards of acceptance

With very few exceptions, the acceptance standards quoted and used by different authorities reflect years of experience and are aimed at a 'safe' situation. Inevitably this sometimes means that the design may not make best or most

Fig. 8.5 *Charpy impact testing*

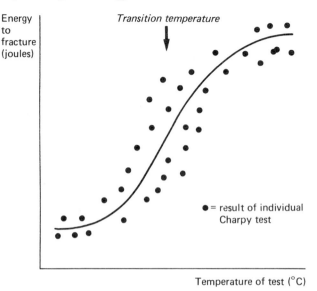

Fig. 8.6 *Effect of test temperature on results of Charpy tests of carbon steel*

efficient use of material, but often this is a small price to pay for confidence in the ability of the structure to withstand service requirements.

A problem which confronts anyone who sets out to define an acceptance standard is the imprecise nature of the measurements. In general engineering it is relatively straightforward to stipulate that the diameter of a shaft should be 100 mm ± 0.1 mm, since the inspector can readily measure this and reject shafts with diameters which are more than 100.1 mm or less than 99.9 mm. On the other hand, how does one specify the maximum amount of porosity that can be allowed in a joint? We could say 'not more than 1 pore per 10 mm of weld', which seems explicit enough, but on reflection there are a number of questions we could ask. What diameter of pore would we tolerate? Is this an average? Would three pores in the middle 10 mm of a 30 mm length be acceptable? Similarly, how do we specify lack of fusion, which has depth as well as length in a weld?

There is no easy answer to these problems and much depends on the experience both of the writer of the standard and of the inspector or surveyor who is interpreting the acceptance criteria. There are many examples of good approaches to be found in standards such as BS 5500, BS 2633 and ASME 9, and these are worth studying.

A more precise approach is offered by the rapidly developing technique of fracture mechanics. If we know the size and location of a planar defect and the stress surrounding it, and we have some measure of the notch ductility of the material, it is possible to predict whether the defect will grow and lead to failure in service. To do this requires a quantitative representation of the notch ductility. Unfortunately, Charpy tests do not give data which can be used in calculations. To get some idea of what we need, we can look back to fig.8.4 and notice that the

Fig. 8.7 CTOD test

difference between ductile and brittle behaviour is typified by the amount of deformation at the tip of the crack. This can be quantified in a CTOD (crack tip opening displacement) test to give data for use in fracture calculations.

In this test (fig.8.7), a slot is cut in the sample and a crack is formed at its root by subjecting the sample to fatigue loading. The sample is cooled to the test temperature and then bent around a support positioned below the crack. Stress is developed at the tip of the crack, and the material there will either deform plastically or fracture with no loss in section. A clip gauge placed at the mouth of the slot measures the amount of displacement.

9 Checking and controlling weld quality

9.1 Non-destructive testing

When weld quality was being discussed in Chapter 8, we recognised the importance of some defects in terms of service behaviour and crack propagation. It follows that if we are to use welded structures with confidence we must have some means of checking whether unacceptable defects are present in the welds. It is also clear that, after testing, the welded joint must be in a fit condition for use. In other words we must use non-destructive testing (NDT) methods.

There are many NDT techniques available to the weld inspector, but they are mostly expensive either in terms of capital equipment or because labour costs are high. Each method has its own limitations and there is no universal testing system. The inspection of welded joints must be carefully planned to ensure that the selected technique is capable of detecting the defects which are of concern

It is also important to specify the acceptance standard before testing is undertaken, since this will dictate the sensitivity required. We can take sensitivity in this context as meaning the smallest defect which can be located with certainty. There is no point in stating that cracks with lengths greater than 1 mm are unacceptable if the test method cannot detect cracks less than 2 mm long – such a situation simply gives a false sense of security. It would also be wrong to say that a weld is free of defects because the inspection technique used did not discover any. A more sensible statement would be that there were no flaws detected.

We must also recognise that NDT of itself cannot tell us if a weld is acceptable or not. All we can report is a list of the types of defects discovered, their size, and their location in the joint. It is for the accepting authority, be it the insurance surveyor, customer's representative, or in-house inspector, to decide if the level of defects is greater than that which the designer would be prepared to tolerate in the service environment of the fabrication.

9.2 Visual inspection

It is a common mistake to think that NDT as applied to welded joints consists exclusively of radiography or ultrasonic flaw detection, which we shall discuss later. Before either of these techniques is used, the experienced inspector will examine the joint visually. Often defects can be discovered by the naked eye and can be rectified or repaired at this stage.

In visual inspection, attention is paid to three aspects:

a) dimensions of the weld (especially in fillet welds, figs 9.1 and 9.2),
b) penetration in joints welded from one side,
c) surface defects.

Fig. 9.1 *Simple fillet-weld gauges*

Fig. 9.2 *Universal weld gauge (Welding Institute):* top *throat measurement;* bottom *leg-length measurement*

9.3 Assisted visual inspection

Even with good lighting, it is often difficult to detect cracks by visual inspection. The effectiveness and indeed the speed of the examination is improved if the presence of a crack is highlighted by the use of either dye-penetrant or magnetic-particle techniques.

Dye-penetrant inspection

If a liquid which has a low surface tension is poured on to the surface of the welded joint, it wets the metal and flows uniformly (fig. 9.3). In particular, it seeps into a crack or cavity. Wiping the surface of the parent metal and weld leaves the liquid in the crack. If an absorbent layer of chalk is now deposited on the clean surface, the liquid is drawn out of the crack. By adding a red dye to the liquid, the location of the crack is indicated by a red stain on the chalk. Alternatively, the liquid may contain a dye which fluoresces when viewed under ultra-violet light.

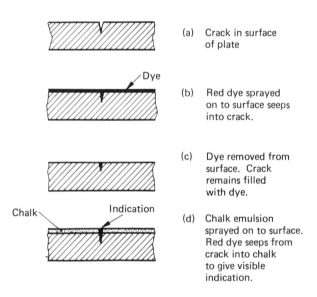

(a) Crack in surface of plate

(b) Red dye sprayed on to surface seeps into crack.

(c) Dye removed from surface. Crack remains filled with dye.

(d) Chalk emulsion sprayed on to surface. Red dye seeps from crack into chalk to give visible indication.

Fig. 9.3 *Principles of dye-penetrant inspection*

Dye-penetrant inspection is very easy to use on the shop floor or on site, since the equipment required is very portable. The dye remover and chalk 'developer' are obtainable in aerosol cans. The choice between chalk developer and fluorescent light is very much a matter of judgement based on local circumstances. Fluorescent dye may give somewhat better definition of the outline of the crack, but it is necessary to view in subdued background light. In addition, the ultra-violet illuminator requires a power supply. Against this, the use of a chalk developer requires more time and, if left, the dye spreads across the surface of the chalk, making the indication more diffuse.

Magnetic-particle inspection

When inspecting ferromagnetic materials such as carbon steel, an alternative method of locating surface cracks is magnetic-particle inspection (MPI). In this technique, a magnetic field is created in the crack and is used to attract iron-oxide particles.

Consider a magnet in the shape of a bar (fig. 9.4). The lines of force within the magnet run from one end of the bar to the other, i.e. from the S to N poles. At the same time a magnetic field exists around the magnet. Iron powder sprinkled on to the bar collects around the poles, where the lines of force are close together. If another magnet is brought into proximity so that its N pole is opposite the S pole of the first magnet, the lines of force flow between the two magnets. When iron powder is sprinkled on to the new arrangement it collects in and around the gap between the N and S poles.

(a) Field around single magnet

(b) Field between N and S poles

Field leaks out to surface

(c) Field around a crack in a bar or plate

Fig. 9.4 *Magnetic fields around bar magnets and a crack*

We can make a crack in a weld in steel behave in the same way as the gap between the magnets by inducing a magnetic field at right angles to it. The lines of force flow across the crack and 'leak' out to the surface. Iron or iron-oxide particles deposited on the weld collect at the point where the magnetic field is leaking, thus indicating the presence of a crack.

Permanent or electromagnets can be used to create a magnetic field in the welded joint. More commonly, a current is passed through the area being examined (fig. 9.5). The detecting fluid or 'ink' is usually particles of magnetic iron oxide suspended in kerosene – the particles may be coated with a fluorescent compound for viewing under ultra-violet light. In cases where there are objections to the use of kerosene, the magnetic particles can be dispersed from a 'puffer' to form a cloud above the surface of the weld. This alternative is particularly useful when inspecting root runs in 'V' joints, since there is no need to clean the groove before depositing the next run of weld metal.

Fig. 9.5 *Magnetic-particle inspection using alternating current*

9.4 Radiography

Visual inspection, with or without the aid of dye-penetrant and magnetic particle techniques, can give positive information only about defects which appear on the surface. Traditionally, radiography has been used to locate defects in the body of the weld, although in recent years ultrasonics has also become recognised as a complimentary method of non destructive testing.

The detection of defects in welds by radiography is based on the ability of X-rays and gamma rays to penetrate materials which are opaque to ordinary white light. Both are electromagnetic rays and have characteristic wave lengths:

X-rays: 2×10^{-12} m to 10^{-9} m
gamma-rays: 10^{-13} m to 2×10^{-12} m

X-rays are produced at will in a suitable vacuum tube, while gamma rays are emitted by radioactive material. In terms of flaw detection they behave in similar ways, and for the following explanation of the principles of radiography we will not distinguish between them. Later in the chapter their specific characteristics will be discussed.

During transmission through the material the X-rays are absorbed. If the material is homogeneous, the amount of absorption is uniform across the area exposed to the X-ray beam. On the other hand, if the material contains, say, a pore of gas, a smaller amount of the rays passing through this point is absorbed and

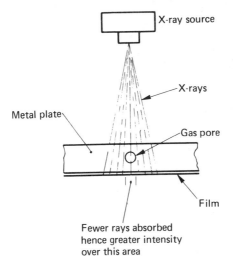

X-ray source

X-rays

Metal plate

Gas pore

Film

Fewer rays absorbed
hence greater intensity
over this area

Fig. 9.6 *Principles of radiography*

there is a variation in the intensity of the emergent beam. This can be readily detected by placing a photographic film on the side of the material opposite to the source of the radiation (fig. 9.6). The film is exposed by the X-rays and its optical density depends on the intensity of the radiation. Thus the area under the gas pore receives more X-rays than the parent material on either side and the photographic emulsion is affected to a greater extent. On a negative film this shows up as a dark spot of the same shape as the pore (fig. 9.7).

The difference in intensities of the emergent rays depends on the relative absorption coefficients of the parent material and of the gas in the pore. The coefficient of absorption is a function of the atomic number of the material and the wavelength of the X-rays. Fortunately, defects in welded joints usually contain either gas, air, or slag, which have appreciably smaller absorption coefficients than the parent metal. Their presence is thus readily detected, since there is a marked variation in density in the film. The only exception to this is

Plate

Weld containing
wormhole porosity

Plate

Fig. 9.7 *Typical radiograph*

provided by tungsten inclusions: this metal is denser than those normally welded and gives a bright spot on the negative film, since more X-rays are absorbed.

Not all defects are detected by radiography – much depends on the orientation of the flaw with respect to the beam (fig. 9.8).

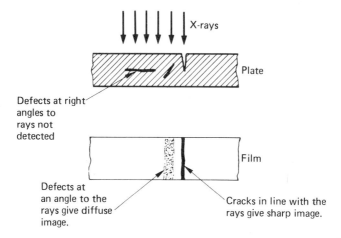

Fig. 9.8 *Orientation of defect with respect to X-rays*

Taking a radiograph

When a weld is to be radiographed, any extraneous matter on the surfaces is first cleaned off to ensure that there are no spurious indications. Sometimes the excess metal and underbead are removed by grinding, to give more uniform conditions.

The film is held in a lightproof cassette which is placed against one side of the joint. The X-ray source is fixed at a suitable distance from the weld on the opposite side of the joint and is arranged so that the beam falls symmetrically on to the surface of the metal. The source is switched on and the joint is exposed to the X-rays for a predetermined period. Exposure times depend on the intensity of the radiation, the wavelength of the rays, and the thickness of the metal being examined – they can range from 10 seconds to 10 minutes for X-rays and 1 minute to 24 hours for gamma rays. After exposure, the cassette is taken to a darkroom where the film is developed and fixed in a similar way to normal photographic film. When it has dried, the film is viewed against a bright light.

One of the advantages of radiography is that the film can be stored as a permanent record of the quality of the weld.

Generation of X-rays

X-rays are generated in a vacuum tube which contains a cathode and an anode (fig. 9.9). The heated cathode emits electrons which are accelerated towards the anode by a high voltage difference of the order of 100 to 400 kV. The quantity of electrons depends on the temperature to which the cathode is heated and is therefore a function of the current flowing in the cathode filament (2 to 10 mA). The anode consists of a tungsten disc (called the anticathode or target) set in a copper block. As the electrons hit the target, their kinetic energy is released with

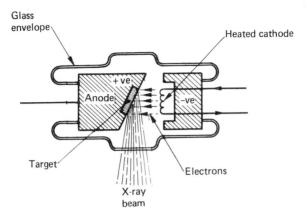

Fig. 9.9 *Simplified construction of an X-ray tube*

the generation of both X-rays and heat. The heat is conducted away by the copper block, which in high-power tubes may be liquid-cooled. The target is at an angle to the beam of electrons, so that the X-rays are emitted through a window towards the work.

Gamma-ray sources

Some elements have unstable atomic nuclei which disintegrate with the passage of time. This disintegration is accompanied by the emission of radiation, part of which is composed of gamma rays. These have high energy and can penetrate metals in the same way as X-rays. They can therefore be used for the examination of welded joints.

For the inspection of welds, 'artificial' radioactive isotopes, which are obtained by fission or irradiation in a nuclear reactor, are used to provide a source of gamma rays. The most popular isotopes are iridium-192, cobalt 60, caesium-137, and caesium-134.

Isotopes are small and, unlike X-ray sources, do not need a power-supply unit. They are readily portable, but must be handled with care and stored in lead-lined containers. The gamma rays are radiated in all directions, a feature which can be of particular value when inspecting pipe welds, as the source can be located in the bore with the film wrapped around the outside of the pipe.

Safety in the use of radiography

X-rays and gamma rays can be extremely harmful to human beings since they damage body cells. Depending on the dosage received, exposure to this form of radiation can result in ailments ranging from burns to leukemia and cancerous growths. The industrial use of radiography is the subject of strict legislation (e.g. *Ionising radiation regulations*, HMSO) designed to protect not only the people taking the radiograph but also those engaged in other operations in the vicinity.

The effects of exposure to radiation are cumulative, and persons working continuously with radiographic equipment are required to wear film badges which measure the amount of radiation they have received over a stated period of time. This is known as the dose rate and, in practice, provisions must be made to keep it below a certain level. In addition, the testing area should be screened to ensure that the radiation does not reach other workers. The screens are preferably lead-lined or are made of lead-bearing brick.

X-rays are usually more manageable since they are highly directional – whenever possible, the tube and work should be arranged so that the beam is pointing at the floor. Gamma rays are given off in all directions from an isotope and pose greater problems in screening. The area in which radiation may exist must be identified by the use of the international warning symbol. After the film has been exposed, an X-ray machine can be switched off and is then safe. On the other hand, a radioactive isotope emits radiation continuously and must be kept in a shielded container when not in use.

9.5 Ultrasonic inspection

Ultrasonic inspection offers a highly sensitive method of detecting flaws inside a metal by observing the way in which high-frequency vibrations are transmitted through the metal. If the material being tested contains a flaw, the vibrations are reflected. The presence of the flaw can be indicated either by noting the reduction in the strength of the transmitted vibrations or by monitoring the reflections.

This type of behaviour is typical of a wide range of vibration frequencies. We are familiar with the reflection of sound waves from a wall or a sheet of metal; this happens at frequencies from 16 Hz, corresponding to the lowest note on an organ, up to the point at which we can no longer hear the sound, somewhere between 15 and 20 kHz depending on the individual person.

Detecting ultrasonic waves

Reflections from an interface still occur at higher frequencies, i.e. in the ultrasonic range, but we cannot recognise the effect with the unaided ear. Instead, we must use some kind of sensor. In the ultrasonic testing of metals, frequencies between 0 5 and 10 MHz are used, and the vibrations are measured by using the piezo-electrical characteristics of a crystal.

If a disc of quartz of uniform thickness is subjected to a pressure, positive and negative charges are produced on opposite faces. When the pressure is removed and a tensile force is applied through the thickness, the charges are reversed. By placing the quartz crystal in the path of the ultrasonic vibrations it is alternately subjected to compression and tension, and the charges on its face alternate between positive and negative. These charges can be fed to a cathode-ray oscilloscope where they can be displayed, thus giving a useful method of detecting the presence of ultrasonic vibrations (fig. 9.10).

Generation of ultrasonic vibrations

We can also use the quartz crystal as a generator of vibrations by reversing the process. If an alternating electric potential is applied to it, the crystal emits ultrasonic vibrations. A problem arises when we try to inject these vibrations into the piece of metal being tested. Simply placing the crystal in contact with the metal surface leaves an air gap, giving an interface which reflects or attenuates (i.e. reduces) the signal. Coating the surface with oil or grease overcomes this by ensuring good contact between the quartz and the metal.

Fig. 9.10 Using a quartz crystal to detect ultrasonic vibrations

Ultrasonic testing of welded joints

There are two ways of using ultrasonic waves in the testing of welded joints: transmission and reflection.

Transmission

Separate transmit and receive probes are placed on opposite sides of the plate. These are connected to a visual display unit (VDU). The magnitude of the received signal is displayed on the cathode ray tube (CRT) built into the VDU. If there are no defects then there is only a slight attenuation of the signal which reaches the receiver (fig. 9.11). A small defect in the path of the ultrasonic beam reflects some of the waves, and the signal detected on the receiver side is reduced. On the other hand, a large defect such as a lamination in the plate causes a complete loss of the transmission signal.

With this method of testing there is no way of determining how deep the defect is below the surface. Major drawbacks are the need to have good access to both sides of the component or section and the ever-present problem of ensuring accurate alignment of the probes. If the probes are not directly opposite each other, the indication will be as if there were a defect present.

Reflection

A combined transmitter/receiver probe is used for this method and access is required to only one surface.

With a *normal probe*, the signal is transmitted at right angles to the surface (fig. 9.12). In the absence of a defect, the ultrasonic waves travel through the thickness of the plate, are reflected from the other surface, and return to the probe. They are displayed on the VDU as a boundary echo. The VDU also records the entry of the waves into the plate as the transmitter signal. The distance between this and the boundary echo is a measure of the thickness of the plate.

When there is a defect in the path of the waves, some of the signal is reflected and, having a shorter distance to travel, appears on the VDU before the boundary echo. The depth of the defect below the surface can be deduced from its position

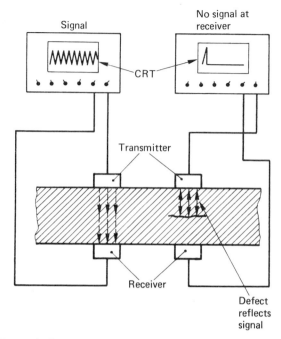

Fig. 9.11 *Transmission technique for detecting flaws*

Fig. 9.12 *Reflection technique using a normal probe which injects pulses of ultrasonic vibrations at right angles to the plate surface*

136

CRT display
with no defect
present

Entry | | Defect
echo

CRT

Weld

Signals reflected from
lack of side-wall fusion

Fig. 9.13 *Reflection technique with an angle probe*

relative to the transmitter and boundary indications. Separate transmit and receive crystals mounted in one probe can be used, but a single crystal transmitting and receiving pulses is probably more common in current practice.

An *angle probe* fitted with a single crystal is normally used to test welded joints (fig. 9.13). A pulse of ultrasonic vibrations is transmitted into the plate at an angle to the surface. This travels to the opposite surface, where it is reflected at an angle equal to the angle of incidence. The vibrations are again reflected from the top surface, and so on. If there is no interface in their path, the vibrations continue until they peter out. On the other hand, a defect reflects some or all of the signal, which may then follow the original path or take a new route back to the

Probe moved backwards and
forwards between these
lines

Fig. 9.14 *Moving the probe to scan the thickness of the weld*

probe, depending on the orientation of the interface with respect to the beam. A defect echo is displayed on the VDU, and again its distance from the transmitter signal will be a measure of the distance travelled by the ultrasonic waves. Knowing the geometry of the system, the exact location of the defect can be calculated. In the examination of a weld (fig. 9.14, previous page), the operator marks two lines parallel to the weld and moves the probe backwards and forwards between these two limits to ensure that the complete thickness of the weld is examined . Scanning of a complete welded joint can be mechanised and the results plotted to give a 'map' of the defects. One such technique is called C-scan. Table 9.1 summarises the methods of non-destructively testing welds.

Table 9.1 *Summary of the methods of non-destructively testing welds*

Method	Defects detected	Advantages	Limitations
Visual	Inaccuracies in size and shape. Surface cracks and porosity, under-cut, overlap, crater faults.	Easy to apply at any stage of fabri-cation and welding. Low cost both in capital and labour.	Does not provide a permanent record. Provides positive information only for surface defects.
Dye-penetrant	Surface cracks which may be missed by naked eye.	Easy to use. No equipment re-quired. Low cost both in materials and labour.	Only surface cracks detected with certainty. No permanent record.
Magnetic-particle	Surface cracks which may be missed by naked eye. May give indication of sub-surface flaws.	Relatively low cost. Portable. Gives clear indication.	Only surface cracks detected with certainty. Can be used only on ferro magnetic metals. Can give spurious indications. No permanent record.
Radiography	Porosity, slag inclusions, cavities, and lack of penetration. Cracks and lack of fusion if correctly orientated with respect to beam	Can be controlled to give reproducible results. Gives permanent record	Expensive equip ment.Strict safety precautions required. Better suited to butt joints not very satisfactory with fillet-welded joints Requires high level of skill in choosing conditions and interpreting results.
Ultrasonics	All sub-surface defects. Laminations.	Very sensitive – can detect defects too small to be discovered by other methods. Equipment is portable. Access required to only one side.	Permanent record is difficult to obtain. Requires high level of skill in interpreting cathode-ray-tube indications.

9.6 Controlling weld quality

The non-destructive tests described in sections 9.2 to 9.5 are of considerable value since, by providing details of any defects which can be located, they give us a degree of confidence in the ability of the welded joint to withstand service conditions. This is all they can do.

As we have already discussed, they will not tell us if the weld is good enough for our purposes; equally important, they will not guarantee that welds will be made which are free of defects.

In other words, we cannot control quality simply by specifying that the welds will be subjected to NDT on completion. Admittedly the knowledge that the welds will be tested may act as an incentive to do better work, but it is far more relevant to establish a system which encourages the production of defect-free welds at all times.

Welding is usually the last of a series of operations involved in fabricating a structure. The quality of the weld reflects what has gone before. This is an aspect of quality planning that is often overlooked, and welders can justifiably claim that on many occasions they are blamed for faults which are the direct result of factors beyond their control. It has even been claimed that one of the welder's tasks is to put right an accumulation of manufacturing defects.

Welding is a corporate activity in the sense that the successful completion of the welding operation, both in terms of quality and economics, is determined by decisions made by a number of people who often do not have a direct contact with the process.

Take as an example the fillet weld in fig 7.4, and assume that it was deposited with a hydrogen-controlled MMA electrode. What factors contributed to the formation of the cracks, and who was responsible? The following is a list of the more significant considerations.

Contributing factor	Area of responsibility
Was the steel a weldable grade?	Designer
Was the plate to specification?	Purchasing
Was the plate correctly identified?	Stores
Who devised the welding procedure?	Client, Welding Engineer
Was the procedure tested and approved?	Welding Engineer
Was a procedure sheet prepared?	Welding Engineer
Was the correct electrode specified?	Design, Welding Engineer
Were the electrodes stored correctly?	Stores
Was the joint assembled accurately?	Fabrication shop
Was the preheat correctly applied?	Welding shop
Were the electrodes dried?	Welder
Was the welding procedure followed?	Welder
Was the correct technique used?	Welder
Was the joint allowed to cool slowly?	Welding shop

In this analysis, the welder is ultimately responsible for only three possible causes of any cracking.

If the welding operation is to be completed satisfactorily, the input of the other areas of responsibility involved in production needs to be coordinated. In the first instance, it is the province of *Quality Assurance* (QA) to provide the environment within which this coordination can be achieved. The management structure, lines of command and responsibility for taking decisions are defined in a Quality Manual which also contains procedures to correct aberrations in the system.

The quality assurance provisions in an organisation are not just confined to welding. They cover the whole design and manufacturing operation. Nationally-approved accreditation agencies issue sets of objectives in accordance with accepted standards, e.g. BS EN ISO 9001–30. Specific quality requirements for welding are contained in a new European Standard: EN 729 Pts: 1 to 4.

Whilst QA provides the framework, monitoring the application to the welding operation is the task of *Quality Control* (QC). The way in which this is done depends on the size of the organisation and the amount of welding involved. In large fabricators there may be a separate QC Department. In smaller concerns inspectors may form part of the welding shop team.

If shop floor inspection is to be effective and consistent it should be fully documented as a Code of Practice. A typical code would include:

- a schedule of inspection tasks;
- a clear statement of acceptance criteria;
- procedures to be adopted if defects in excess of permitted levels are identified.

Staff allocated to the inspection of welds should have technical knowledge to make informed judgements.

Not only should they be versed in the company's products, they should hold qualifications issued by an approved national body.

9.7 Approving the welding procedure

The way in which the weld is deposited is one of the key factors in achieving quality. For many applications, especially with low-strength structural steels, the welding operation is tolerant of such details as the size of weld runs and the sequence used to deposit them. By contrast, when welding creep-resistant steels, the properties of the completed joint are critically dependent on the preheat level, heat input (size of weld run) and run sequence. In applications such as this, before any attempt is made to weld the finished product, there is a need to confirm that the procedure used to make the joint does in fact produce the required results. In this context the procedure includes:

- materials in terms of specification, composition and surface condition
- related pre-weld operations such as edge preparation and joint assembly
- preheat temperature and methods of achieving it
- values for the welding variables, and run sequence
- acceptance criteria and method of assessment
- post-weld treatments

In reality all welds are made to a procedure, but not all are documented and tested prior to welding. Often material and joint type are specified but it is left to the welder to devise the procedure based on experience. More commonly the term weld procedure is reserved for those situations where there is a need to meet a specific requirement and there are strict instructions which the welder must follow. The purpose of the procedure may be to:

- achieve specific properties, eg impact strength
- ensure freedom from critical defects such as h.a.z. and weld metal cracks
- enforce QC procedures
- standardise costs in repetitive work
- control production schedules

The specific requirements of a weld procedure for a particular application may be:

- defined in a published standard or code of practice
- specified by a client
- dictated by production requirements

To demonstrate the validity of the procedure it can be tested under supervision by an independent authority and subjected to approval.

Broadly there are three groups of people who could be interested in the approval of a weld procedure. Firstly, the customer may have a member of staff or may appoint a representative who will be responsible for ensuring that the welding conforms with the requirements of the contract. In this context, the customer can be a private company or a government or nationalised organisation. Secondly, an official body may be involved in cases where the public safety is of concern when the fabrication is in service. Finally, an insurance company or a client may want the procedure to be approved by one of its surveyors before agreeing to insure.

When an approval is required, the weld procedure is first specified in terms of

a) welding process,
b) parent metal,
c) consumables – specification, care, etc.,
d) shop or site weld,
e) preparation (cleaning, edge preparation, assembly, etc.),
f) position of welding,
g) number and arrangement of runs,
h) electrical conditions,
i) gas shielding,
j) heat input,
k) preheat and interpass temperature,
l) back-gouging,
m) post-weld heat treatment,

This list is not exhaustive but indicates the detail with which a procedure should be specified.

A sample weld is prepared using the procedure and is tested according to an agreed code. Any of the interested parties discussed above may have its own standards of acceptance or may agree with the manufacturer to accept a British Standard or other published national standard.

In general terms, the standard will call for the test plate to be examined by NDT, probably stipulating that a radiograph should be taken for record purposes,

Fig. 9.15 Method of cutting test-pieces from procedure-approval plate

and for test-pieces to be cut from the plate for destructive testing (fig.9.15). Usually a transverse tensile test is made to demonstrate that the failure is in the parent plate, thus confirming that the correct electrode material has been selected. With aluminium and its alloys, the joint efficiency would be determined.

Bend tests are also a normal requirement, but their exact form will depend on the standard being used. The aim of the bend test is to place one part of the welded joint into tension with the object of confirming that the metal has sufficient ductility to bend round a specified radius and assessing if any defects present will initiate cracks. Three forms of test are currently used (fig. 9.16):

a) face bend, in which the top surface of the weld is in tension;
b) root or reverse bend, in which either the root of the weld or the reverse face of the weld is in tension;
c) side bend, in which a vertical slice is cut transverse to the weld so that the cross-section of the joint is in tension.

142

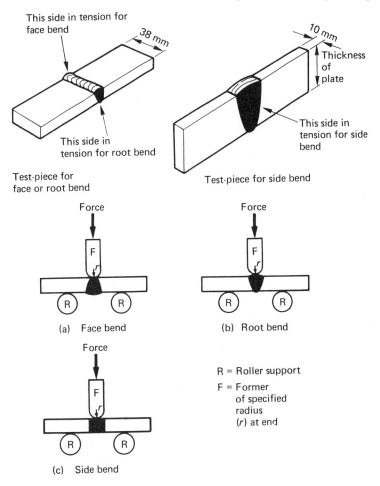

Fig. 9.16 *Three types of bend test*

In each case the test strip, which is prepared to the dimensions and finish specified in the standard, is placed across two rollers and a former is pressed against the opposite side to that which is being tested. The test-piece is forced downwards between the rollers, so that the lower surface is in tension, until the sides are parallel. The surface is then examined for cracks and tears.

In the UK the key standard for approval of welding procedures is BS EN288 'Specification and approval of procedures for fusion welding of metallic materials' which is based on the agreed standard for use in the European Union. This supersedes BS4870 and is divided into four parts:

Part 1 – General rules for fusion welding
Part 2 – Welding procedure specification for arc welding
Part 3 – Welding procedure tests for the arc welding of steels
Part 4 – Welding procedure tests for the arc welding of aluminium and its alloys

Widely used American standards for weld procedures are:

> ASME VIII – Boiler and Pressure Vessel Code
> AWS D1.1 – Structural Welding Code

Details of the approved procedure are usually put on a weld procedure sheet (Table 9.2), which is distributed to personnel concerned with its implementation.

Table 9.2 *Welding procedure sheet*

Typical Procedure Sheet for Manual Welding

Order No: 24021/AX/94	Drawing No: WS/24021/0824-04	Sheet No: WP1294	Joint Ref: WP1294/1 to 3

PROCESS TIG	MATERIAL Aluminium magnesium alloy 5083		

JOINT PREPARATION: WELDING SEQUENCE

RG=1.5mm $^{+0}_{-1}$ RF=1mm $^{+.5}_{-0}$

CLEANING Vapour degrease	GOUGING METHOD N.A.

WELDING CONDITIONS

ELECTRODE	CONSUMABLES	Pass No.	Electrode Size (mm)	Filler diam. (mm)	A.C. or Polarity	Amps	Volts	Wire Feed speed m/min	Position
Specification 1% Zirconiated	Filler Wire BS2901: 5356	1	2.4	2.4	A.C.	90	–	–	F
		2	4.0	3.2	A.C.	210	–	–	F
Drying Method N.A.	Shielding Gas Argon	3	4.0	3.2	A.C.	210	–	–	F
	Flow Rate 7 l/min								
	Flux N.A.								

Preheat None	INSPECTION		
Interpass Temperature None	Standard Customer specification RT/56/1994	Method	
		Visual	✓
		Radiographic	✓
Post Heat None	Inspecting Authority Customer's representative	Ultrasonic	☐
		Dye Penetrant	✓
		Magnetic Particle	☐

Notes Root run (pass 1) to be inspected by dye penetrant before depositing passes 2 and 3				

Originator: GWT	Checked: P.J.	Approved: JF Smith	Date of Issue: 24/9/94	Returned to file:

9.8 Approving the welder

In addition to approving the procedure, the welders must be checked to confirm that they can in fact produce welds to the quality represented by the test plate which was accepted by the inspecting authority. For this purpose the welder pre pares another test plate similar to that used for the procedure approval. This is tested, perhaps not quite as exhaustively, and if it is satisfactory the welder is approved for work on the fabrication.

In the UK welder approval tests are specified in BS EN287, which supersedes BS 4871. The standard is in two parts:

Part 1 – Steels
Part 2 – Aluminium and aluminium alloys

The standard makes provision for a 'job-knowledge' examination to test the welder's understanding of safety, consumables, weld defects, approvals and the welding process. This test is not mandatory in all countries but can be requested by clients if it is thought necessary. Welders, especially those working in the oil and energy industries, may be tested to an American standard, ASME IX, which is part of the Boiler and Pressure Vessel Code referred to under weld procedure.

In some cases, rather than approving the welder for each job, it may be agreed that a general qualification will cover a range of work, provided the standard required is not higher than that achieved by the welder in the test.

Welder approval, which is often referred to as 'coding' of welders, can only tell us that under the conditions of the test the welder achieved a particular standard. It demonstrates the level of competence we can expect from the welder, but it would be wrong to assume that the same quality will always be achieved in welds made under production conditions – it simply provides a basis on which quality control of the actual welding operation can be established.

10 Shrinkage and distortion in fusion welding

10.1 Thermal expansion and contraction

Thermal expansion and contraction of metals are such familar phenomena in engineering that we often forget that without them many tasks could not be performed as readily, nor would many mechanisms function properly. Shrink fitting of tyres on to wheels and sleeves on to shafts is practised regularly under closely controlled conditions. Thermal cut-outs rely on metal strips expanding in a predictable fashion. Conversely, the same phenomena can cause problems, as when overheated parts expand and foul their bearings. The physical laws governing the expansion and contraction of metals are well known and ample data are available to enable many applications to be analysed mathematically.

Since fusion welding involves a heating-and-cooling cycle, it is inevitable that thermally induced dimensional changes occur which have a profound effect on the behaviour of the joint during the joining operation. Unfortunately, the sequence of thermal events involved in welding is far from simple and is not readily amenable to mathematical analysis. On the other hand, it is possible to describe the contraction of a welded joint in a qualitative manner and to ascribe to the different stages empirical data established by observations made over a period of many years.

In practice we are principally concerned with the shrinkage that takes place during cooling after the heat source has passed along the joint line, although, of course, any restrained expansion resulting from heating before fusion may cause undesirable buckling, for example in thin sheets.

Shrinkage during welding is made up of three components:

a) contraction of liquid metal,
b) change of volume on solidification,
c) contraction of solid metal.

10.2 Solidification shrinkage

In practice it is difficult to separate the contraction of liquid metal and the change of volume on solidification, since melting and solidification in welding are progressing at the same time. Looking at a transverse vertical section through a welded joint during the solidification cycle (fig. 10.1), the two stages can be seen to interact. As the solidification front progresses upwards towards the centre-line, the solid metal occupies a smaller space than the liquid metal it replaces, i.e. its density is higher. The molten metal is also contracting, with the result that the surface of the pool should recede below the original level. However, at the same time, further molten metal is being fed into the area by melting of the parent metal

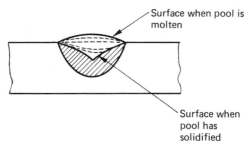

Surface when pool is
molten

Surface when
pool has
solidified

Fig. 10.1 *Shrinkage during solidification*

at the leading edge of the weld pool and by the addition of the melted electrode material. Thus, although appreciable shrinkage is occurring during cooling and solidification from the liquid state, the surface of the weld will not normally show evidence of this unless insufficient added metal has been available to compensate for the reduction in volume.

The end or termination of a weld run gives an interesting example of the importance of weld-metal feeding. If the heat source is suddenly removed, solidification proceeds as normal, but, since no further melting is taking place, the level of the surface falls and a crater is produced. In practice, the welder adds sufficient electrode or filler metal to build up the surface, possibly at the same time gradually reducing the heat input. This technique, known as crater filling, is of considerable importance in welding, since cracking can often be associated with an incorrectly or inadequately filled crater.

10.3 Contraction of solid metal

After welding, the weld metal is still hot and heat continues to flow into the parent metal until the whole joint has cooled to room temperature. During this period of cooling, the weld metal contracts both along the length of the joint and at right angles to it. This means that its volume should become smaller, but in fact this is prevented from happening by the restraining influence of the parent metal to which it is bonded. It is not easy to visualise this effect with an actual weld, because a number of events are happening at the same time, but we can illustrate the final results by considering longitudinal shrinkage in a weld.

Longitudinal shrinkage

Some idea of the decrease in length of the weld can be obtained by using the equation

$$l_1 = l_0 (1 - \alpha\Delta\theta) \text{ or } l_1 = l_0 - l_0\alpha\Delta\theta$$

where

l_0 = original length
l_1 = length after cooling through $\Delta\theta°$C
α = coefficient of linear expansion
$\Delta\theta$ = temperature change (°C)

The linear shrinkage or contraction is therefore equal to $l_0 \alpha \Delta \theta$

For low-carbon steel, the following data would apply:

$\alpha = 14.3 \times 10^{-6}/°C$

$\Delta \theta$ = melting temperature – room temperature
$= 1500°C - 20°C$
$= 1480°C$

Suppose the weld is 1 metre long:

$l_0 \alpha \Delta \theta = 1000 \text{ mm} \times 14.3 \times 10^{-6}/°C \times 1480°C$
$= 21.2 \text{ mm}$

This figure is only an approximation, since the value of α does not remain constant over the temperature range we quoted, but it does indicate that the contraction along the length of the weld should be appreciable. In practice the measured contraction is significantly less, being of the order of 1mm per metre length of weld. We must therefore examine why this should be so.

The first and most obvious factor is that the hot weld element is attached, i.e. bonded, to the cold plates on either side and is not free to shrink. The plates can be considered as being more or less rigid, so that, when the weld metal tries to contract, it is plastically deformed by being held to the original length. This means that tensile forces are being set up in the weld region, which are reacted by compressive forces in the plates (fig. 10.2).

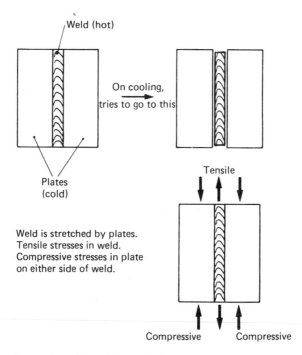

Fig. 10.2 *Deformation of a weld-metal element during cooling*

If these plates were perfectly rigid, when the cooling cycle had finished the welded joint would be the same length as the plates were originally. The compressive stresses induced, however, are of considerable magnitude and are in excess of the compressive yield stress of the parent metal. The result is that the plates themselves are plastically deformed, so reducing the overall length of the joint and thus accounting for the 1 mm/m shrinkage quoted above. This is appreciably less than the amount of deformation which occurs in the weld element, principally because the yield stress of the relatively cold plate is higher than that of the hot weld metal. Some idea of the magnitude of the forces generated during shrinkage in welded joints may be obtained from the fact that, in low-carbon steel at the temperature involved, a compressive force of about 150 to 170 N/mm^2 is required to achieve a compressive strain of about 1×10^{-3} mm/mm (i.e. a shortening of 1 mm per metre length of joint).

Transverse shrinkage

Similar considerations apply when we look at shrinkage transverse to the weld, where the contracting weld metal tries to pull the plates towards the centre-line of the joint and as a result the whole joint area is in transverse tension. Again we have a situation where, because the hot weld metal has a lower yield stress than the cold plates, deformation first takes place in the weld but, at a later stage of cooling, as the relative yield stresses become more equal, some yielding of the parent material occurs and the overall width of the welded plates is reduced.

Strictly, the amount of transverse shrinkage which takes place depends on the total volume of weld metal, but as a general rule, for a given plate thickness, the overall reduction in width transverse to the joint at any point is related directly to the cross-sectional area of the weld. Similarly, as we would expect, the total shrinkage increases with the thickness of the plate, since the weld area is greater. It is possible to state the relationship in a general way:

transverse shrinkage = kAt
where k = an empirical factor with a value between 0.1 and 0.17
A = cross-sectional area of weld
t = thickness of plate

This formula can be used to predict the shrinkage that will occur in a butt joint (fig. 10.3) and has been found to give good correlation with practical observations. In the case of a single-V butt joint the calculation can be simplified, since the ratio A/t is equal to the average width (fig. 10.4) and the formula is reduced to:

transverse shrinkage = $k \times$ average width of weld

It should be noted that for a double-V weld the average width is not zero but is the value for *one* of the Vs.

Transverse shrinkage = $0.1 \times A$
$A = a + b + c$
$\quad = \frac{1}{2} \times 5 \times (12 + 3) + (3 \times 12) + (\frac{1}{2} \times 12 \times 12)$
$\quad = 145.5$ mm
∴ transverse shrinkage = 0.1×145.5
$$\frac{}{12}$$
$$= 1.21 \text{ mm}$$

Fig. 10.3 *Estimation of transverse shrinkage in a T butt joint*

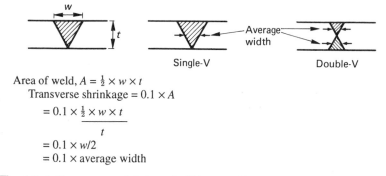

Single-V   Double-V

Area of weld, $A = \frac{1}{2} \times w \times t$
Transverse shrinkage = $0.1 \times A$
$\quad = 0.1 \times \frac{1}{2} \times w \times t$
$$\frac{}{t}$$
$\quad = 0.1 \times w/2$
$\quad = 0.1 \times$ average width

Fig. 10.4 *Transverse shrinkage in V butt welds*

Angular distortion and longitudinal bowing

Taking both longitudinal and transverse shrinkage, based on what has been said above, the final shape of two plates welded together with a butt joint should be as shown in fig. 10.5(a). In practice, however, such a simple treatment does not apply, principally because the shrinkage is not distributed uniformly about the neutral axis of the plate and the weld cools progressively, not all at one time.

If we look at a butt weld made with a 60° included-angle preparation, it is immediately apparent that the weld width at the top of the joint is appreciably greater than at the root. Since the shrinkage is proportional to the length of metal cooling, there is a greater contraction at the top of the weld. If the plates are free to move, as they mostly are in fabricating operations, they will rotate with respect to each other. This movement is known as angular distortion (fig 10.5(b)) and poses problems for the fabricator since the plates and joint must be flattened if the finished product is to be acceptable. Attempts must be made, therefore, to reduce

150

(a)   Changes in shape resulting from
      shrinkage which is uniform throughout the thickness

(b) Asymmetrical shrinkage tends to produce
    distortion.

Fig. 10.5 *Change in shape and dirnensions in butt-welded plate*

the amount of angular distortion to a minimum. Clamps can be used to restrain the movement of the plates or sheets making up the joint, but this is frequently not possible and attention has to be turned to devising a suitable weld procedure which aims to balance the amount of shrinkage about the neutral axis. In general, two approaches can be used: weld both sides of the joint or use an edge preparation which gives a more uniform width of weld through the thickness of the plate (fig. 10.6).

In the direction of welding, asymmetrical shrinkage shows up as longitudinal bowing (fig. 10.7). This is a cumulative effect which builds up as the heating-and-cooling cycle progresses along the joint, and some control can be achieved by welding short lengths on a planned or random-distribution basis (fig. 10.8). Welding both sides of the joint corrects some of the bowing, but can occasionally be accompanied by local buckling.

Angular distortion and longitudinal bowing are also observed in joints made with fillet welds (figs 10.9 and 10.10). Angular distortion is readily seen in this case as a reduction of the angle between the plates and is greatest for the first weld. Although the second weld, placed on the other side of the joint,tends to pull the web plate back into line, the amount of angular rotation will be smaller. With experience, the joint can be set up with the web plate arranged so that the first angle is greater than 90° and thus ends up with the web and flange at right angles. Even so, warping in the flange plate cannot be ignored.

(a)

(b)

(c)

a) Double-V joints balance the shrinkage so that more or less equal amount of contraction occur on each side of the neutral axis. This gives less angular distortion than a single 'V'. b) It is difficult to get a completely flat joint with a symmetrical double 'V' as the first weld run always produces more angular rotation than subsequent runs; hence an asymmetrical preparation is used so that the larger amount of weld metal on the second side pulls back the distortion which occurred when the first side was welded. c) Alternatively, a single-U preparation with nearly parallel sides can be used. This gives an approach to a uniform weld width through the section.

Fig. 10.6 *Edge preparation designed to reduce angular distortion*

Longitudinal
distortion

Direction of welding

Fig. 10.7 *Longitudinal bowing or distortion in a butt joint*

Fig. 10.8 *Sequences for welding short lengths of a joint to reduce longitudinal bowing*

Fig. 10.9 *Longitudinal bowing in a fillet-welded T joint*

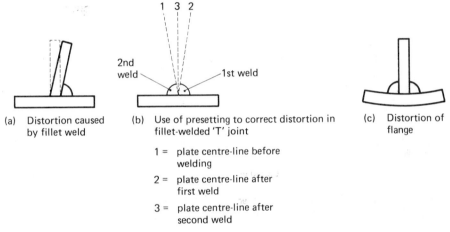

(a) Distortion caused by fillet weld

(b) Use of presetting to correct distortion in fillet-welded 'T' joint

2nd weld

1st weld

1 3 2

(c) Distortion of flange

1 = plate centre-line before welding

2 = plate centre-line after first weld

3 = plate centre-line after second weld

Fig. 10.10 *Distortion in fillet welding of T joints*

Effect of heat distribution

Finally, in our consideration of shrinkage and distortion we must not ignore the importance of heat input. As we have seen in Chapters 2 and 3, the heat from the weld pool during solidification flows into the plate adjacent to the fusion boundary. The width of metal heated to above room temperature is greater than that of the fused zone, and the picture used above of a hot weld-metal element between cold plates is an over-simplification. The heat flowing into the plates establishes a temperature gradient which falls from the melting point at the fusion boundary to ambient temperature at some point in the parent metal remote from the weld.

The width of the heated band is directly proportional to the heat input in joules per mm length of weld and is therefore dependent on the process being used. It follows that the amount of distortion and shrinkage will also vary from one welding process to another. If the heat source moves slowly along the joint, the heat spreads into the plate and the width of hot metal which must contract is greater. The effect is less noticeable in thick plate but in sheet material, say 2 mm thick, the differences are most marked. The MIG/MAG system, with its fast

153

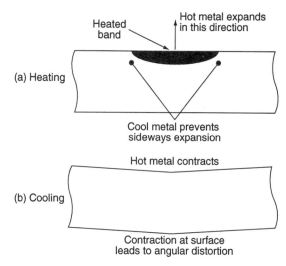

Heated band

Hot metal expands in this direction

(a) Heating

Cool metal prevents sideways expansion

Hot metal contracts

(b) Cooling

Contraction at surface leads to angular distortion

Fig. 10.11 Effect on contracting and expanding in heated metal

speed of travel, gives a narrow heat band compared with the spread in oxy-acetylene welding, and it is possible to arrange the manual processes in ascending level of distortion, i.e. MIG/MAG, MMA, TIG, and oxy-acetylene welding.

10.4 Flame straightening

Straightening steel plates and sections which have been distorted during welding often involves the use of large capacity presses. A novel approach to the problem lies in using the physical principles which created the distortion in the first place. If a narrow band of metal at the surface of a steel plate is heated to about 1,000°C, the metal expands (fig 10.11). The cold plate on either side constrains lateral expansion and the metal is forced to expand at right angles to the surface. On cooling, the metal contracts and creates forces which distort the plate in the same way as a cooling weld produces transverse distortion.

Applied to a welded T joint in steel, the distortion resulting from heating a band of metal can be used to straighten a distorted flange (fig.10.12). A large oxy-acetylene torch is used to heat the surface. The tip of the inner cone of the flame is positioned at the surface of the plate and the travel speed is adjusted to give a heated band width equal to the plate thickness. If the distortion is severe, it may be necessary to repeat the process placing hot bands on either side of the first pass. Alternatively, two torches can be used simultaneously. As it is not possible to measure the temperature of the heated band, the aim is to raise the surface of the steel to a cherry red colour. Successful use of flame straightening relies heavily on experience gained by trial and error.

10.5. Residual stresses

Solving the problems of distortion control during welding and determining shrinkage allowances for design purposes are of such importance in fabrication that it is easy to overlook the fact that they are the products of plastic deformation

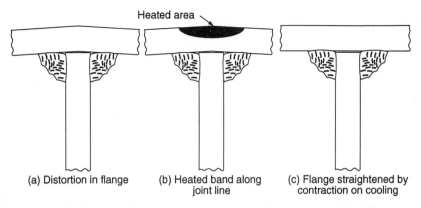

Heated area

(a) Distortion in flange   (b) Heated band along   (c) Flange straightened by
                                  joint line   contraction on cooling

Fig. 10.12 *Using flame straightening to correct distortion in welded flange*

resulting from stresses induced by contraction in the joint. As long as these
stresses are above the yield point of the metal at the prevailing temperature, they
continue to produce permanent deformation, but in so doing they are relieved and
fall to yield-stress level. They then cease to cause further distortion. But, if at this
point we could release the weld from the plate by cutting along the joint line, it
would shrink further because, even when distortion has stopped, the weld
contains an elastic strain equivalent to the yield stress. Visualise the completed
joint as an element of weld metal being stretched elastically between two plates.

The stresses left in the joint after welding are referred to as residual stresses.
From our discussion of shrinkage and distortion, it can be seen that there will be
both longitudinal and transverse tension. In the case of the longitudinal stresses,
the weld itself and some of the plate which has been heated are at or near yield-
stress level (fig. 10.13). Moving out into the plate from the heat-affected zone, the
stresses first fall to zero. Beyond this there is a region of compressive stress.

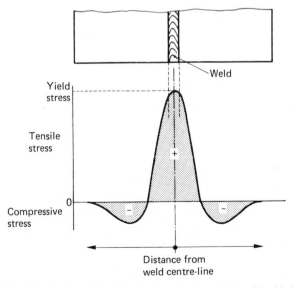

Fig. 10.13 *Distribution of residual stresses in a butt-welded joint*

155

It must be emphasised that all fusion welds which have not been subjected to post-weld treatments – in other words, the vast majority of welded joints – contain residual stresses. Procedures developed to minimise distortion may well alter the distribution of the residual stresses but do not eliminate them or even reduce their peak level. Having said this, since we cannot avoid the formation of residual stresses, it is appropriate to ask if we are worried by their presence. As with so many engineering situations the answer is not a simple yes or no. There are numerous applications where the existence of residual stresses would have little or no influence on the service behaviour of the joint – storage tanks, building frames, low-pressure pipework, and domestic equipment all provide examples of situations where the joints can be used in the as-welded condition without detriment.

If the service requirements do indicate that the residual stresses are undesirable, the designer must take them into account when selecting materials and deciding upon a safe working stress. This approach can be seen in the design of ships, where the combination of low temperatures and residual stress could lead to a type of failure known as brittle fracture. The designer selects a material which is not susceptible to this mode of failure even at the low temperatures which may be experienced during the working life of the ship; the presence of residual stresses is then not important. Similarly, in many structures subjected to loads which fluctuate during service – for example, bridges, earth-moving equipment, and cranes – the designer recognises the existence of residual stresses by choosing a working-stress range which takes account of the role these stresses play in the formation and propagation of fatigue cracks. There are, however, some specific applications where it is essential to reduce the level of residual stresses in the welded joint. With pressure vessels, because of the risk of a catastrophic failure by brittle fracture, stress-relieving is often a statutory or insurance requirement. Again, some metals in certain environments corrode rapidly in the presence of tensile stress, i.e. stress corrosion will occur. In these cases, a joint in the as-welded condition containing residual stresses suffers excessive attack; this is retarded if the joint is stress-relieved. Finally, when machining welded components, removing layers of metal near the joint may disturb the balance between the tensile and compressive residual stresses and further deformation or warping can occur. This can make it difficult to hold critical machining tolerances and it may be desirable in these circumstances to stress-relieve to achieve dimensional stability.

Stress-relieving

Various methods are available to reduce the level of residual stresses in welded joints. Heat treatment, overloading, and vibratory treatment can all be used, but the most common method is based on a controlled heating-and cooling cycle, i.e. thermal stress relief. This technique makes use of the fact that the yield stress of a metal decreases as the temperature is raised. If a welded joint is heated to, say, 600°C, the residual tensile stress, which was equivalent to the yield stress at room temperature, is in excess of the yield stress of the metal at 600°C. Localised plastic deformation occurs, and the tensile stresses are reduced. At the same time, the compressive stresses which were in equilibrium with the tensile stresses are also reduced, to restore the equilibrium. In stress-relieving practice, the tempera-

Table 10.1 *Example of stress-relieving temperatures for fusion-welded pressure vessels*

Type of steel	Stress-relieving temperature (°C)
Low-carbon	580
Carbon-manganese	580
Carbon-$\frac{1}{2}$% molybdenum	650
1% chromium – $\frac{1}{2}$% molybdenum	650
2$\frac{1}{4}$% chromium – 1% molybdenum	690
5% chromium–$\frac{1}{2}$% molybdenum	725

ture is raised until the yield stress has fallen to a low value at which residual stresses can no longer be supported. This clearly depends on the metal being treated, since the relationship between yield stress and temperature is critically influenced by alloy content. And this is characterised by the typical stress-releiving termperatures in Table 10.1.

If thermal treatment is to give a satisfactory reduction of residual-stress levels, it is important that differential expansion and contraction must not occur, otherwise new residual stresses will be induced. The heating and cooling must be carefully controlled so that the temperature is uniform throughout the component, and special furnaces equipped with comprehensive temperature-control systems have been designed for this purpose. In these furnaces the whole of the component or fabrication is heated, thus easing the problem of avoiding temperature gradients. Localised heating for stress relief is usually not recommended,

Fig. 10.14 *Typical specification for temperature distribution during local stress relief of welded butt joints in pipe*

especially with joints in flat plates since there is always the risk of creating further stresses. In this connection, pipe welding poses particular problems. Stress relieving might often be desirable to reduce corrosion problems, but it would be impracticable to heat-treat a complete pipework installation. Local stress relief of pipe joints *in situ* is, therefore, allowed by some authorities, provided that the temperature distribution is controlled. This is usually achieved by specifying the minimum temperature at the joint line and at some specific point remote from the weld – a typical example is shown in fig. 10 .14.

11 Mechanising arc welding

11.1 Factors affecting welding speed

Welding is used in so many widely differing industries that it is difficult to assess the relative usage of the various processes. However, it is probably not too inaccurate to suggest that the manual processes discussed in Chapters 5 and 6 account for about 90% of all the fusion welding of plates, pipes, sections, castings, and forgings carried out in the UK. To some extent this reflects the fact that much of the work undertaken benefits from the versatility of the manual systems. Relatively short runs, constant changes in the position of welding, movement across the shop floor from one fabrication to another, and a mix of different types of work or joint all encourage the use of manual welding, since the welder's skill can be deployed to the best advantage in coping with the varying requirements. On the other hand, some fabrications, especially where long joints are involved, could be designed to take advantage of higher speeds of welding or greater deposition rates than those obtainable with manual techniques.

High welding speeds involve more than just travelling faster along the joint. If a sound bond is to be achieved in the weld, the minimum heat input for fusion must always be exceeded. Therefore, the rate of heat input to the arc must be increased as the travel speed is raised. This follows from the relationship given in Chapter 3:

$$\text{heat input to weld pool (kJ/mm)} = \frac{\text{current (A)} \times \text{arc voltage (V)} \times 60}{\text{weld travel speed (mm/mm)} \times 1000}$$

To achieve high welding speed we must use a high current, as we cannot make large changes in arc voltage. Similarly, to deposit significantly larger weld runs than those produced in the normal operating range for manual welding (100 to 350 A) the current must be increased to between 500 and 1000 A to give more rapid electrode melting (fig. 11.1). Currents of this magnitude are not easily managed manually, and we must look at some form of mechanisation which takes from the welder the tasks of controlling the arc length, feeding the electrode, and traversing the joint.

11.2 Mechanised TIG welding

Probably the easiest system to mechanise is TIG welding, especially for use on thin sheet where filler metal is not required. In essence, all that is needed is a method of moving the TIG torch along the joint line at a fixed height above the work, thus keeping the arc length constant.

Fig. 11.1 *Relationship between welding current and electrode melting rate in arc welding*

This assumes that the work is flat, at least along the joint line, to within the limits allowed for variations in arc length. More often than not when sheet is being welded by the TIG process, the abutting edges lift and distort under the influence of the heat from the arc. In manual welding, the operator can allow for this and make adjustments to keep the arc length constant. For mechanised welding, the sheets must be clamped to prevent the edges lifting, and special jigs of the type shown in (fig. 11.2) need to be used. These are fitted with a backing bar against which the sheets are clamped. The bar is usually made of copper, since this helps to conduct the heat away from the weld area. The clamping force is provided by a series of 'fingers', each about 75 to 100 mm wide and fitted with copper tips which, by conducting heat away from the weld, help to keep the heated width as small as possible, thus minimising distortion.

Where arc-length control is critical, it is possible to hold the torch in a clamp which can be raised or lowered in response to commands from a servo-control system linked to the arc voltage.

11.3 Gravity welding

In general, it is not easy to mechanise manual metal arc welding. The short length of the electrodes and the difficulties of controlling the arc length have hampered the development and exploitation of suitable systems. One technique which has been successfully used although not widely is gravity welding (fig. 11.3). The electrode is clamped in a holder which can slide up and down a bar or rod held at a predetermined angle to the joint.

The improvement in deposition rate over manual welding is small, but, by using one operator to look after four or five welders, overall output can be higher. Although simple, gravity welding is not widely used (fig. 11.4).

Fig. 11.2 *Longitudinal clamping system for mechanised TIG welding*

Fig. 11.3 *Gravity welding*

Fig. 11.4 Arc-length control in gravity welding

11.4 Continuous-electrode systems

Undoubtedly the most effective way to mechanise arc welding is to use a continuous electrode coupled with a suitable method of arc-length control.

We have already seen a continuous electrode used with a self-adjusting arc in MIG/MAG welding (page 60), and many mechanised systems are based on this principle, but it will be remembered that self-adjustment works well only when fast electrode feed rates are used, such as would be the case with small-diameter electrodes. There are considerable advantages to be gained, however, by using larger diameter wires; for example, higher currents and deposition rates can be achieved at slower wire feed speeds, thus simplifying the construction of the wire-drive unit. In some ways this situation is then similar to that encountered in MMA, where the welder controls the arc length by adjusting the rate at which the

Fig. 11.5 Control panel for a submerged-arc welder

electrode is moved towards the weld pool. With mechanised welding, the electrode is fed into the arc area by means of a motor and a drive-roll system. This is the method adopted for Submerged Arc Welding (SAW – page 34). The arc operates beneath a protective layer of granulated flux and is, therefore, not visible. Arc length control is achieved by monitoring the arc voltage and adjusting the wire feed speed to correct any variations from the norm.

The voltage, current, and travel speed are set on the control panel (fig. 11.5). Current is supplied by a drooping-characteristic power-supply unit (fig. 11.6) which keeps the current at the predetermined value. The motor feeds the electrode at a speed which gives the desired arc length at the current being supplied by the power unit. The voltage corresponding to this arc length is used as a reference against which the actual arc voltage measured during the welding operation is compared. If the arc length increases, the voltage rises. The comparator sends a command signal for the motor control to increase the speed in proportion to the difference between the arc and the reference voltages. Since the burn-off rate of the electrode is determined mainly by the current and varies only slightly with voltage, the electrode is now moving faster than it is melting and the tip approaches the weld pool, thus reducing the arc length to its original value. At the same time, the arc voltage falls back to the reference voltage and, as the arc approaches its original length, the electrode feed rate slows down until it also returns to the preset value. Table 11.1 summarises the two systems used in mechanised arc welding.

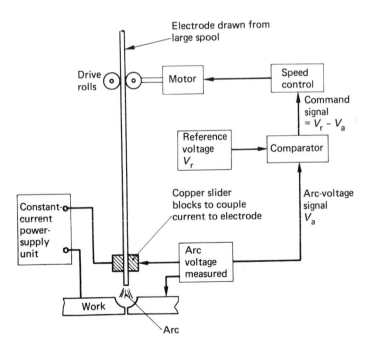

Fig. 11.6 *Typical layout of a controlled-arc system for consumable-electrode welding*

163

Table 11.1 *Mechanised welding arc-length control systems*

System	Electrode dia. (mm)	Motor drive	Power supply
Self-adjusting arc	0.8 to 3.0	Constant speed at preset value to determine current	Flat characteristics with voltage selector
Controlled arc	3.0 to 6.0	Variable speed; servo control based on arc voltage	Drooping charac-teristics with current selector

11.5 Consumables in submerged-arc welding

The consumables in SAW welding are the wire and the flux. Two types of wire are used: solid and tubular (flux-cored). The solid wires are used with either a neutral or a basic flux.

Solid bare wires

Bare wire electrodes for mechanised systems are similar to those used for MIG/MAG welding in that they contain alloying elements needed for strength and de-oxidation. The alloy content is adjusted to compensate for losses in the arc and by absorption into the flux. For example, to achieve 1.0% manganese in a steel deposit, the wire may need to contain as much as 1.8% of this element.

Steel wires range from 0.8mm to 6.0mm diameter and are copper coated to minimise rusting in storage and to ensure good electrical contact with the copper sliders or blocks which transfer the welding current from the power supply unit. Uncoated wires are available if the presence of small amounts of copper in the weld impairs mechanical properties.

Tubular wires and flux-cored electrodes

Tubular wires for submerged arc welding are similar to the flux-cored electrodes discussed on page 95. They offer the advantage of greater composition flexibility in manufacture. It is also claimed that they offer higher rates of deposition than solid wires as the smaller metal cross-section results in higher current densities and, hence, more rapid heating.

Tubular wires are particularly useful in the deposition of hard-wearing surface layers. These deposits contain relatively large amounts of alloying elements such as manganese, tungsten, cobalt and chromium. Steels containing appreciable percentages of these alloys are difficult to manufacture as wires. It is much more economic and technically desirable to use a low carbon steel for the tubular shell and to pack the core with alloying elements. As the electrode melts, the alloys mix with the steel from the tube and a metal of suitable composition is deposited onto the surface of the component.

Fluxes

The flux in SAW welding performs a number of functions besides providing an atmosphere for the arc. It cleans the surface of the weld pool and influences the surface profile of the finished weld. For most work on low carbon and lower strength structural steels, the flux is neutral. Although it may absorb some alloying elements, thereby depleting the weld metal, it does not enter into reactions with the molten pool and, therefore, is not involved in de-oxidation. Basic fluxes are used for welding higher strength steels where hydrogen control is needed and where it would be advantageous to introduce alloying elements in addition to those contained in the wire.

Fluxes are manufactured in two forms, both granulated: fused or agglomerated. *Fused fluxes* are made by melting the components and casting the liquid so that it breaks into small glass-like particles. These are crushed to a powder and graded for size. Fused fluxes are recognisable by their shiny surfaces. They produce a slag which is easily removed to reveal a smooth weld surface.

Fused fluxes are invariably neutral as it is difficult to add and retain alloying elements and deoxidants in manufacture during the melting process. Basic fluxes are produced by agglomeration which involves bonding at much lower temperatures than those used in making fused fluxes. *Agglomerated fluxes* have a rough appearance similar to powder used in washing machines, although the granule size is appreciably larger. They offer the advantage of alloying and are essential for welding higher strength steels, but they require more careful storage than fused fluxes if the hydrogen content of the weld is to be controlled to low levels. The slag may be difficult to remove, and the weld surface is not as smooth as with fused fluxes.

11.6 Practical welding systems

The various combinations of electrode type and shielding system give a number of possible mechanised processes, as shown in Table 11.2.

Submerged-arc welding is the mechanised process most commonly found in fabricating and welding shops. It is used to weld pressure-vessel shells, box- and I-section girders for bridges, stiffened panels for ships, longitudinal butt joints in the manufacture of pipes, and many other products where long straight joints are used (see page 35). Welding currents can range from as low as 200 A up to more

Table 11.2 Mechanised welding systems

Electrode type	Shielding method		
	Flux	Gas	None
Solid bare-wire electrode	Submerged-arc welding	MIG welding	
Flux-cored electrode (Tubular wire)	Hard-facing systems	Flux-cored arc welding (MIG)	Self-shielded flux-cored welding

165

than 1000 A. In practice the range tends to be more restricted, and, for most applications, the current is between 450 A and 850 A. Below 450 A, submerged arc welding offers little advantage over MIG/MAG welding or the flux-cored system with or without a gas shield.

A number of alternative traversing mechanisms can be used. The main requirement is that the method selected should be capable of moving the complete welding head – i.e. wire feed, electrode spool, and control box – at a stable controllable speed along the joint line. Some typical examples from current practice are:

a) hand-operated trolley, sometimes called a deck welder;
b) tractor unit, electrically driven along a portable track;
c) motorised drive along a horizontal beam;
d) screw-fed saddle on a lathe-type bed;
e) hydraulic ram.

Alternatively, the welding head can remain fixed and the work be moved beneath it. This technique is frequently used when welding large-diameter pipes or shells.

With submerged-arc welding, a strip electrode which gives wide deposits on flat surfaces can be used where a large surface is to be covered with a dissimilar metal (fig. 11.8). For example, the inside of a steel chemical vessel can be covered in this way with a stainless-steel weld deposit to improve corrosion-resistance.

Fig. 11.8 *Using a strip electrode in submerged-arc welding to deposit surface layers on to plates*

Edge preparations for submerged arc welding

Welding at currents up to 1000 A not only enables welding speeds to be increased, but this practice also leads to deeper penetration. This has a significant effect on the dimensions and type of edge preparation for joints which are to be welded by SAW. A square edge can be used up to 10 mm thickness. The included angle in both a single-V and double-V can be reduced to 40° (20° bevel). The double-V can be used up to 30 mm thickness, at which point it is preferable to change to a U type preparation (page 51). The reduction in the amount of electrode metal needed to fill a joint which results from these changes in edge profile is lost if an incorrect design is used. For example, using an edge preparation intended for MMA welding (60° included angle and small root face) can result in a doubling of the amount of electrode used thus losing some of the economic advantages of SAW. A factor which must not be overlooked in choosing an edge preparation for SAW is the difficulty experienced in controlling root penetration when using high currents. This is discussed later in this chapter (fig. 11.11).

A development in SAW which results directly from the ability to reconsider the design of edge preparation is *narrow gap welding* (fig 11.9). With this technique the joint preparation is almost parallel sided with a gap of between 15mm and 20mm. The root gap can be closed with a backing strip, or a modified J-preparation can be used. Although the weld can be deposited in one run, it is more common practice to complete it in two passes. With such a narrow gap there is the danger that the electrode wire touches the side-walls of the joint and arcing can be erratic. The success of the technique depends on keeping the wire central; this is achieved by use of a guide which can be moved between the faces of the joint.

11.7 Advantages in using mechanised arc welding

From our discussion of the ways in which arc welding can be mechanised, five clear advantages for its use can be identified:

a) It offers the ability to use high welding currents, which means that faster welding speeds can be employed.

Fig. 11.9 Edge preparation for narrow gap welding

b) It is possible to weld longer lengths without stopping.
c) Because the traverse speed, arc voltage, arc current, and wire-feed speed are all controlled, the welds can be deposited with more consistency than in manual welding, i.e. there is better reproducibility. This can be a considerable attraction where quality is to be controlled by monitoring welding parameters.

It should be noted that the benefits of reproducibility can only be achieved if the input to the welding operation is consistent. In other words, the fit-up must be accurate: joint gaps uniform, bevel angles consistent, and plates aligned correctly. At the same time, the full implication of reproducibility must be recognised. If the correct welding conditions and fit-up are chosen, the welds can be reproducibly good; but if, say, the heat input is set wrongly at the start, lack of fusion may be reproduced along the length of the weld.

d) With high welding speeds it is possible to achieve smaller heated-band widths, accompanied by a reduction in the amount of shrinkage and distortion.
e) As the current is increased, not only does the deposition rate become larger but the weld also penetrates deeper into the plate. This can be a distinct advantage in the welding of T joints (fig. 11.10).

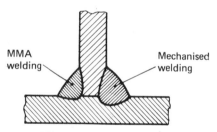

MMA welding

Mechanised welding

Fig. 11.10 *Comparison of penetration in welds of the same leg length deposited by MMA and mechanised welding*

11.8 Limitations of mechanised welding

At high currents, the weld pool is large and we have difficulty in controlling the molten metal unless it is held in place by gravity. Admittedly fillet welds can be deposited in the horizontal-vertical position, but there is always a tendency for the metal to run on to the horizontal plate and form an overlap. We can restrict the current and/or weld size in horizontal-vertical welding to avoid this defect, but the degree of success depends on the type of flux, if any, which is used. It follows that, for the best results from mechanised arc welding, the work must be manipulated so that welds are deposited in the flat position.

The high currents used in mechanised welding raise problems in depositing root runs. The deep penetration achieved makes it difficult to control the melt-through, and, without some form of support, uniform-penetration beads are not usually a practical proposition except with mechanised TIG welding. This means that, when butt welding a plate, it is preferable to weld from one side first, turn the plate over, gouge a groove along the joint line, and then complete the joint from the second side. Where it is essential that the welding is from one side only, various methods of backing may be used to support the weld, some of which are illustrated in fig. 11.11.

(a) Permanent strip
     of same metal as
     work

(b) Removeable copper
     backing bar

(c) Flux backing
     supported in metal
     channel

(d) Adhesive-tape
     backing

Fig. 11.11 *Methods of backing joints during mechanised welding*

11.9 Electro-slag welding

In all the mechanised processes considered so far, we have adopted the conventional practice of using an arc to establish a weld pool and then moving the heat source along the joint line to give progressive fusion and solidification. Where the thickness of the plate rules out the possibility of completing the weld in one pass, plate-edge preparation and multi-pass techniques are used. A completely different approach to the welding of thicker plates is to place the joint vertical with a gap of, say, 25 to 50 mm between the square edges of the two plates and to fill the gap with molten metal (fig. 11.12). Water-cooled copper shoes must be provided on both sides of the plates, to prevent the molten metal from running out of the joint, and these shoes are moved up the joint as the gap is filled.

In many ways the system is similar to continuous casting, with the important exception that the 'cast' metal fuses to the plates making up two sides of the 'mould'. To achieve fusion of the plates, which is necessary for good bonding, the heat source must operate in the gap itself. An arc can be used for this purpose, and a flux-cored electrode with a gas shield (electro-gas welding) has been used successfully on plates from 12 to 50 mm thick.

It is more common, however, to use the heat generated by the passage of a current through a molten slag bath (fig. 11.13). When solid, welding fluxes are non-conductive. On the other hand, when the flux or slag is molten, it conducts current but has a significant electrical resistance. Appreciable amounts of heat are generated when a high current flows in a molten slag bath positioned between the two plates, and both the parent metal and the electrode will be melted. Slag-bath heating is used in electro slag welding and has been successfully applied to the welding of plates from 20 mm up to 350 mm thick.

The welding operation is begun on a piece of scrap plate attached to the bottom end of the joint (fig. 11.14). An arc is struck between the electrode and the starter plate. Granular flux is added and is melted by the arc. As soon as the gap is filled with molten flux, the arc is automatically extinguished and resistance heating begins. As the electrode melts, it fills the gap with metal which mixes with the fused plate material, and the surface of the weld pool rises up the joint. Once equilibrium conditions have been established, solidification at the bottom of the pool proceeds at a steady rate, and the weld pool is of constant depth. Some of the

Fig. 11.12 *Schematic diagram of the arrangement for electro-slag welding*

Fig. 11.13 *Diagrammatic section through an electro-slag welding system*

170

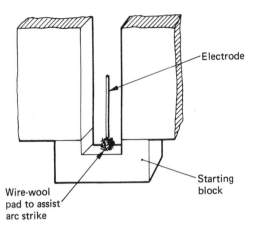

Fig. 11.14 *Starting an electro-slag weld*

slag coats the faces of the copper shoes and the surface of the solidified weld, and it is necessary to add small amounts of granulated flux from time to time to compensate for this loss.

Two electrode-feeding systems are in use at present. In electro-slag welding, one or more electrodes are introduced into the gap just above the surface of the slag bath, and the complete wire-feed unit is moved up the joint as the gap is progressively filled with weld metal (fig. 11.15). The alternative is to feed the electrode down a long tube positioned in the gap before welding begins. With this variant, known as consumable-guide welding, the wire-feed unit is positioned at

Fig. 11.15 *Electro-slag wire-feed unit*

Fig. 11.16 *Wire feed for consumable-guide welding*

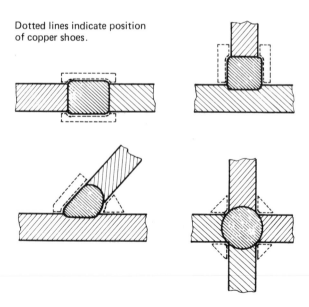

Fig. 11.17 *Typical configuration for electro slag-weld joints*

172

the top of the joint and is not moved during the welding operation (fig. 11.16). The latter system is mechanically simpler and is suitable for welding shorter lengths of joint. The guide, which can take various forms, is melted into the joint.

The first 25 mm or so of the weld when the arc was being used to establish a slag bath is invariably defective, containing oxides and porosity. The starter plate is, therefore, designed so that it can be cut off, leaving sound metal at the start of the actual joint. It follows that, once an electro-slag or consumable guide weld has been started, it should not be stopped until the top of the joint has been reached, as any attempt to restart the system will leave a volume of defective weld metal in the joint.

The weld produced by the electro-slag process tends to be slightly barrel shaped in cross-section (fig. 11.17) and is relatively free from both transverse and longitudinal distortion. Shrinkage will still occur, of course, but, since distortion is negligible, this can be predicted and allowed for. This feature of electro-slag welding has made it particularly attractive in shipbuilding, where vertical joints in the hulls of large tankers have been successfully welded. The process is also used in the fabrication of thick-walled pressure vessels, but here there may be a need to post-weld heat-treat the structure to restore the notch ductility which may have been lost in the joint area due to the very slow cooling cycles in the heat-affected zone. Typical welding speeds are 1 to 3 m/h, depending on thickness These may at first sight seem to be very low, but, when it is realised that they are for completed welds on plates in excess of 30 mm thick, electro slag welding is seen as one of the fastest processes available to the fabricator.

11.10 Robots for welding

Mechanised welding is at its best on straightline joints, whether butt or T. On some equipments it is possible to steer the welding head so that it follows gentle curves. We can also move the head up and down to allow for changes in the vertical plane. Both of these operations involve the use of templates or optical guiding systems. Sensing devices may also be incorporated to check where the head is in relation to the joint line.

If there are a number of changes of direction along the joint, mechanised welding becomes difficult to set up and control. Frequently the answer is to stay with manual welding and provide the welder with a manipulator or other aids designed to increase the amount of time spent welding as opposed to setting up, tacking, moving around the component, and so on.

Robots offer an alternative to both manual and mechanised welding.

What is a robot?

Robots take many forms but are essentially mechanised devices which can repeat a sequence of operations with a high degree of consistency. They may be used for a simple operation such as palletising or machine-loading, or they may be programmed to perform complex assembly tasks.

In a welding shop, a robot is a manipulating tool which moves the welding head along the joint line but at the same time maintains the correct relationship between the welding gun and the work. This means altering the gun or torch angle, keeping the nozzle-to-plate distance constant, and introducing a weaving

With polar coordinates, the
position of point A is
defined by the length of
line *R* from the origin and
the angle *θ*.

Fig. 11.18 *Polar-coordinate robot*

movement when needed. The robot can do this only if it is given instructions, so
there must be a facility to program the complete welding sequence, including
control of the welding parameters. Ideally, there should be some form of
feedback, so that the robot can check that it is welding in the right place.

Types of robot

There are three principal types of robot used for welding operations. They are
distinguished by the way the arm moves to position the welding torch or gun in
relation to the joint line.

Polar-coordinate machines

These are similar to the gun-and-turret mounting on an army tank (fig. 11.18).
The vertical column ('turret') can rotate about its axis, and the arm can move in
or out. At the same time, the end of the arm can be moved up or down by pivoting
at the junction of the arm and the column, just like a gun in a tank.

Cartesian-coordinate machines

These position the gun by using three movements: left – right, up – down, and in
– out (figs 11.19 and 11.20). In geometric terms these are the x, y, and z directions.
Sometimes these machines are called 'rectangular-coordinate machines'.

With Cartesian coordinates, the
position of point A is defined
by measurements x, y, and z.

Fig. 11.19 *Cartesian-coordinate robot*

174

Fig. 11.20 Typical coordinate welding robot

Jointed-arm machines

These are closest in operation to the human frame. They have a rotating turret (the human waist) and several joints in the arm which may replicate the shoulder and elbow (figs 11.21 and 11.22). This configuration makes a jointed-arm machine the most flexible of robots, but it requires the most complex control system.

All machines can have a wrist joint attached to the end of the arm, to allow the welding gun to be correctly positioned.

Degrees of freedom

Looking at these different robots, we can readily see that there are a number of independent movements. Some of them give rectilinear motion, i.e. in a straight line. Others are rotational in character.

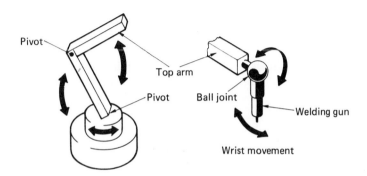

Fig. 11.21 Jointed-arm robot

175

Fig. 11.22 *Typical jointed-arm robot*

Each of the three machines described above has three basic axes. The addition of a wrist movement adds two more axes of rotation giving five in all, which is the minimum required for most welding operations. We can extend the capability of a robot either by moving it bodily along a track or by holding the work in a manipulator which may add one, two, or three axes of motion.

Seam tracking

So far, robots have been mainly used with three welding processes: TIG and MIG/MAG (which we have already discussed in Chapter 6) and resistance welding (which we looked at briefly on page 36, and in Chapter 12).

Resistance welding is particularly well suited to robotic applications, as here the robot's prime task is to position the gun with the electrodes in alignment. The spot weld is completed with the robot stationary.

With both TIG and MIG/MAG welding, the robot moves while welding. The pattern of movement can be complex. At first sight the branch connection in fig. 11.23 seems to be straightforward, but closer inspection shows that the welding gun not only moves along the joint line but also points in different directions. The axis of the gun for a number of positions is indicated by the arrows. This type of movement is not too difficult with a TIG torch. The principal requirements are to position the tip of the electrode along the line of the joint and to maintain the arc length within ± 0.5 mm.

However, considerable difficulty may be experienced if we need to make filler-metal additions with TIG welding. The wire must be introduced into the leading edge of the weld pool and must be along the centre line of the weld. This introduces yet another movement which must be incorporated into the wrist action. Welding into corners poses particular problems, as it becomes impossible to accommodate the wire-feed tube ahead of the weld pool.

Fig. 11.23 *Changes in orientation of the gun axis when welding a branch connection*

MIG/MAG is the arc process most commonly used with robots. It is particularly easy to adapt MIG/MAG welding to robotics, as the arc length is controlled from the power-supply unit and small variations in nozzle-to-plate distance can be tolerated. It is also possible to angle the torch when welding into corners, thus avoiding the problems associated with wire-feeding in TIG welding However, provision does have to be made in MIG/MAG welding to ensure that the wire-feed hose can move freely, and attention needs to be paid to the problems caused by spatter blocking the nozzle and restricting the gas flow. These become particularly important when operating towards the outer limits of the working

Fig. 11.24 *Control system for a welding robot*

177

envelope of the robot. (The working envelope represents the furthest points which can be reached by the end of the arm.) When the arm is fully extended, it may not be possible to position the torch so that the correct welding angle or weaving motion can be maintained.

Controlling a welding robot

The robot must be told what to do. It needs instructions on where to start and stop, what path to follow, and how the torch must be manipulated. To some extent separate from the operation of the robot but nevertheless an integral part of the operation, the welding parameters must be controlled and varied as necessary. These instructions are stored in a memory unit (fig. 11.24).

The memory can be programmed in a number of ways. In most current applications of welding robots, the memory is instructed via a command unit called the teach pendant. This typically has a collection of keys, rather like a large calculator, and it forms the interface between human operator and robot (fig. 11.25). The simplest set of instructions identifies points where a weld is needed. Resistance spot welding is a good example of this type of operation. The robot is driven by the operator to the weld location. This position is recorded, and the robot is driven to the next location, which is also logged in the memory. The procedure is repeated until the positions of all the welds have been noted. The instructions are completed by adding the welding commands for each point. The program is stored in the memory as a series of numbers. When the robot is in operation, this data is supplied to the control unit, which responds by issuing instructions to the various moving parts and to the welding unit.

When a robot is being used for resistance welding, we are usually not too worried about the precise path it takes between spot welds. With arc welding, the situation is different (fig. 11.26). The start and stop positions for the weld run can be fixed in the same way as for resistance welding, but the trajectory or path between these

Fig. 11.25 *Teach pendant for a jointed-arm robot*

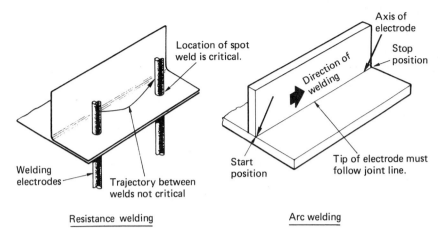

Fig. 11.26 Trajectories in resistance welding and arc welding

two points must be precisely defined as it controls where the weld is deposited. The path can be a straight line or one of a number of predetermined curves.

If the joint line is not easily defined in mathematical terms, e.g. a straight line or an arc of a circle, or has a number of changes of direction, the robot can be driven along the path it is to follow when welding. The coordinates are read at short intervals and again stored in the memory. If there are sudden or frequent changes of direction, allowance has to be made for the possibility of the robot overshooting.

As an alternative to using the teach pendant as a means of inputting data to the robot memory, a welder's movements can be automatically recorded. This is of particular value where the gun has to be manipulated, for example to produce a variable weaving motion. In a case like this, it is convenient to let an operator weld the joint using a gun fitted with a number of sensors which monitor all the critical movements. The data from these sensors is stored in the memory and later used to instruct the robot to produce a weld which copies the one made by the welder.

The above methods of giving instructions are called 'on-line' programming and reflect current practice with most arc-welding applications. Although they give good results, the robot cannot be used for production while it is being programmed, and this can be a lengthy operation. A more satisfactory method would be 'off-line' programming, where the instructions can be written away from the robot and stored on a tape or disk ready to be fed into the memory when needed for production. At present, off-line programming of welding robots is handicapped by the lack of a standard language for the instructions and the need to fine-tune the program on the robot to accommodate inaccuracies in set-up.

Robots – yes or no?

The decision whether or not to install robots is markedly influenced by factors remote from the technological aspects which we have been discussing. A robot system represents a large capital investment, two to three times the cost of the robot itself, which must be evaluated. It requires changes both in management structures and work-force practices. The effects are not confined just to the welding shop – practices in associated design, fabrication, and inspection operations need to be modified to respond to the new demands made by the robots. Staff must be retrained, and production will inevitably be interrupted.

Against all this, the use of welding robots offers significant advantages. Although there may not be increases in welding speed, output per station can be much higher – typically two to four times that of a manual station. Robots can be worked almost continuously throughout a shift, so there is less idle time than with manual operations. Also, unlike human beings, the robot can work 24 hours a day. A major attraction is the greater consistency of the product – this can be an important factor in quality assurance.

The successful introduction of robots depends on a number of technical factors. For example:

a) Joint accuracy must be better than that usually tolerated for manual welding. The welder can accommodate variations in root gap or face by adjusting such things as travel speed, arc length, electrode angle, and weave pattern. The robot is not able to respond so readily – if at all – and would probably give variable penetration or even burn through.
b) Component design must allow for access by the robot arm. Can the gun be brought to the joint with the correct angle and/or nozzle-to-plate distance?
c) If a jig is needed for assembly, it must not interfere with the movement of the robot arm.
d) Distortion control often involves extra movements which can complicate the programming of the robot.
e) Planned maintenance is absolutely essential and may well be novel in a traditional welding shop which uses only manual welding.

12 Resistance welding

We use a whole range of equipment fabricated from sheet metal. Cars, washing machines, refrigerators, radiator panels, and storage cupboards are but a few of the items which are part of modern life. That they are available at reasonable cost to many people is the direct result of developments in engineering techniques for rapid production of identical units. A major factor has been the ability to make joints quickly with minimum distortion. A brief look at a mass-produced sheet-metal component highlights a dependence on lap joints. These are attractive as they can be secured in many ways, some of which were mentioned in Chapter 1. Pop rivets, self-tapping screws, soldering, and adhesive bonding are all used. The manufacturer thus has choice and can concentrate on production-engineering, but, where loads are to be transmitted, resistance welding is often the first choice.

12.1 Basis of resistance welding

Resistance welding depends on the heating effect of a current flowing through an interface between two overlapping sheets (fig. 12.1). The interface offers a

Fig. 12.1 Principles of resistance spot welding

resistance to the flow of the current, and the energy expended is converted to heat. Applying Ohm's law, the voltage (V) required for a current flow (I) is given by V = IR, where R is the resistance of the interface. The total energy for a current flow lasting t seconds is expressed as

$$H = IVt = I(IR)t = I^2Rt \text{ joules}$$

The heat so developed is concentrated into the localised area where the sheets are brought into intimate contact as a result of pressure applied by the electrodes. As the current continues to flow, the temperature rises until melting occurs at the interface, forming a weld pool. If the current is now stopped, the joint area cools and solidification takes place under pressure. The resultant weld nugget provides a localised bond between the sheets which can transmit a load through the joint.

The strength of the weld depends on the cross-sectional area of the nugget at the joint line. In tensile tests, welds with small diameters shear along the interface (fig. 12.2). As the diameter is increased, the load at which such a failure occurs also gets larger. A point is reached where the mode of failure changes and fracture takes place in one of the sheets around the periphery of the weld, leaving the nugget or slug attached to the other sheet.

Small-diameter spot welds shear along interface.

Large-diameter spot welds tear around slug and through sheet.

Fig. 12.2 *Failure of spot welds subjected to load*

Spot-welding sequence

A typical machine for resistance spot welding (fig. 2.40) contains a transformer which reduces the mains voltage to the working voltage (4 to 25 V), a means of controlling the current, and a timer, in addition to a mechanical system for applying force to the electrodes.

Welding sequences can vary, but the simplest (fig. 12.3) would be as follows:

a) The joint is placed between the electrodes. Care must be taken to align the work with the electrodes, to ensure that the weld is in the correct position. Jigs

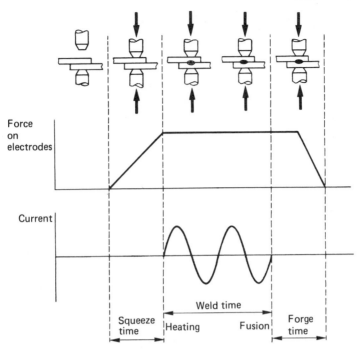

Fig. 12.3 *Typical spot-welding cycle*

are frequently necessary for this purpose if the high throughput rates of which resistance spot welding is capable are to be achieved.

b) The weld sequence is started, usually by operating a foot switch or a pedal.

c) The electrodes close, gripping the work and bringing the sheets into contact.

d) The pressure is increased to an optimum level over a period known as the squeeze time.

e) The current is switched on and is allowed to flow for a pre-set period during which melting occurs and the weld grows to the required size. In general, weld times are appreciably less than one second and are quoted in cycles – one cycle equals 1/50 second, i.e. the inverse of the frequency of the standard a.c. supply in the UK. As an example, for sheet steel (0.6 to 1 mm thick) the aim would be to use weld times between 5 and 20 cycles.

f) The pressure on the electrodes is maintained for a hold or forging time while the weld solidifies.

g) The pressure is finally released, the electrodes are moved apart, and the work is removed from the machine.

Features of a resistance spot weld

The weld which is produced by the sequence described contains three distinctive features (fig. 12.4). The central region, where the interface has been melted, has a cast structure typical of fusion welds, with columnar grains which meet at the line of the original interface. Surrounding the weld nugget is a heat-affected zone, showing that the parent metal has undergone a heating-and-cooling cycle. The outer surfaces of the sheets show indentations resulting from the pressure of

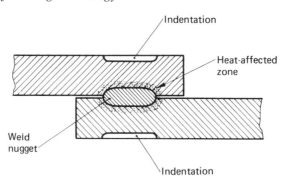

Fig. 12.4 *Features of a spot weld*

the electrode tips. The reduction in thickness of the sheet at this point should not be more than 10% under normal conditions.

Controlling the welding current

The easiest way of altering the current is to adjust the voltage across the electrodes.

According to Ohm's law, $I = V/R$. As R is fixed by the force applied to the electrodes and the surface condition of the metal, any change in V is reflected by a corresponding alteration in the current I. The voltage can be adjusted by changing the turns ratio of the transformer, and most resistance-welding machines are fitted with primary taps. Design considerations preclude the use of a large number of taps, and this method of adjustment can only give a coarse selection of current. In the simplest machines, where the current is switched on and off by an electromagnetic contactor in the main input line, this is the only means available to control current, and any finer setting of heat input must be achieved by adjusting weld time.

Where precise control of the welding sequence is needed, a more accurate method of switching must be used. In Chapter 4 we discussed the use of thyristors in power-supply units for arc welding and we noted that, in addition to their role in rectifying the alternating current, they could also be made to switch the current on at a selected point in each half cycle. A thyristor will not conduct electricity until it has been fired by applying a voltage to its control grid. There is virtually no delay in allowing current to flow in response to the control signal, so the thyristor offers very fast switching.

We must not forget that the thyristor still rectifies the current even when used as a switch. An alternating current is used in resistance welding, so it is necessary to have two sets of thyristors acting in opposite directions. In fig. 12.5, thyristor A looks after the positive half cycles, while B conducts only when the current is negative. By using a timer with a high degree of accuracy to supply the control voltage, weld times can be set precisely.

In older units, ignitrons are used to control current flow (fig. 12.6). These are glass tubes filled with mercury vapour which conduct electricity only when a voltage is applied to the igniter.

A very important facility which follows from the use of thyristors for switching is the ability to control the level of current in the weld circuit. This is

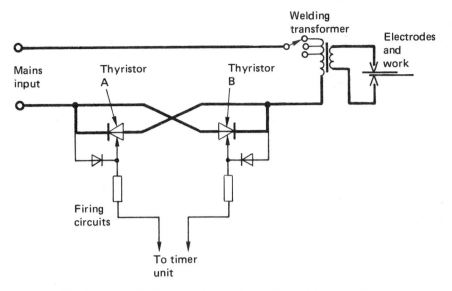

Fig. 12.5 *Thyristors installed in the primary circuit of a resistance welder to control current*

Fig. 12.6 *Main components of an ignitron tube*

frequently called 'heat' control. The power in the weld sequence, and hence the total amount of heat developed, is proportional to the sum of the areas under the current waveform (fig. 12.7). If the start of each half cycle is delayed, a chopped sine wave is produced in which the area under the curve is reduced. The amount of power supplied is related to the proportion of the half cycle during which the current is flowing. Control of this can be achieved by adjusting the length of time between the beginning of each half cycle and the point at which the thyristor is

CURRENT IN PRIMARY

Sine wave shows supply to ignitions.

x = points at which ignitrons are fired

Current flows in primary winding of transformer during shaded periods.

CURRENT IN SECONDARY

Current flows through interface only during periods shown shaded.

Phase shift increased from 1 to 2 to 3
Heat input decreased from 1 to 2 to 3

Fig. 12.7 *Effect of phase shift on current flow through the interface*

fired. This technique is known as phase shifting. The amount of control which can be exercised in this way is limited, as there is a risk that the weld zone will cool between pulses if the delay time is long, and only a small proportion of each half cycle is used. However, by combining phase shifting with a limited number of primary tappings on the transformer, control over the complete current range of the machine can be made available.

Machine rating

It is not very helpful to classify or rate machines by quoting the current flowing in the secondary circuit, as we do not normally know the voltage. At the same time, measuring currents of the order of 2000 to 40 000 A flowing for only fractions of a second poses considerable problems and requires expensive equipment. A more meaningful approach is to quote the maximum amount of power available for conversion to heat at the interface of the sheets being welded. This is commonly expressed in kVA, thus a machine rated at 50 kVA draws 113 A from a 440 V mains supply. Ignoring losses in the transformer and connections, this amount of power is available at the electrodes. The current it produces in the secondary circuit depends, of course, on the voltage provided by a given tapping and on the interface resistance.

There will in fact be some losses in the system. These cause the temperature of the transformer to rise and, if current is drawn for long periods, the insulation

186

could be damaged. To some extent this risk can be minimised by water cooling, but, to give a greater margin of safety, the kVA rating quoted for a machine specifies the power that can be drawn for, say, 30 seconds out of each minute, thus allowing a cooling period.

Current shunting in spot welds

In practical welding situations, a further factor which affects the choice of current level is the spacing between neighbouring spot welds. When the first weld is made, all the current flows from one electrode to the other by the shortest and most direct route, i.e. along the extended axis of the electrode. With subsequent welds, the metallic bridge provided by the first spot weld offers an alternative low-resistance path and some of the current is diverted or shunted along this route (fig. 12.8). The total current is shared between the new site and the existing spot weld. The relative proportions depend mainly on the distance between the two points at the interface. Invariably the second weld is smaller for the same nominal current.

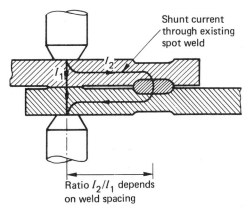

Shunt current through existing spot weld

Ratio I_2/I_1 depends on weld spacing

Fig. 12.8 Current shunting by an adjacent spot weld

Resistance welding high-thermal-conductivity metals

The single-phase a.c. supplies considered so far work well for metals such as low-carbon steel, stainless steel, and nickel alloys. Heat can be developed rapidly at the interface, and melting can be achieved before significant amounts of heat have flowed into the surrounding areas. Metals such as aluminium which have a high thermal conductivity and a low body resistance require very large currents to compensate for the heat losses into the sheets. With thin material, this requirement can be met by using large-capacity machines rated at about 100 kVA. Where the problem is not readily solved by increasing the transformer size, three-phase transformers are used in conjunction with banks of thyristors to provide a suitable power level in the secondary circuit.

D.C. systems

When welding very thin sheet or wire-to-sheet joints, for example in electronic equipment, even the shortest times and lowest currents available from an a.c. machine lead to excessive melting. An alternative is to use a short pulse of d.c. in capacitor spot-welders, also known as stored-energy machines. These units incorporate a capacitor which is charged from a rectifier; they are small and suitable for bench mounting. When the capacitor is connected to the electrode in the welding head, it discharges a high amperage pulse of d.c. through the interface. The duration of the pulse is significantly shorter than one half cycle of 50 Hz, i.e. less than one hundredth of a second. The high current rapidly melts the metal at the interface and the short duration of the pulse ensures that the extent of the melting is constrained. Risk of damage to electronic components is also minimised as there is little spread of heat from the weld area.

The amount of heat developed in the weld is determined by the level of charge stored in the capacitor. Reproducibility with simple stored-energy or capacitor spot-welders is adequate for most applications, but better control of the energy levels and the welding sequence is achievable with transistorised units.

12.2 Practical resistance-welding systems

The principles of fusion by resistance heating which we have discussed in the preceding pages can be used in a variety of ways to weld lap joints in sheet material. However, in terms of industrial practice, only three systems have found wide application:

a) spot welding
b) seam welding
c) projection welding.

Resistance spot welding

In many respects, resistance spot welding is the simplest of the three systems. The current flow is concentrated into a small area at the interface, and the result is a circular region of bonding.

The machines used for spot welding can be either stationary or portable. Stationary or pedestal welders range from small units for welding thin sheet to large machines having high-current outputs. They differ principally in the way in which the force is applied to the electrodes. With simple small machines a foot-operated pedal is linked to a pivoted arm which holds the upper electrode (fig. 12.9). Springs and weights are used to assist the operator and to increase the force

Fig. 12.9 Simplest form of foot-operated resistance spot welder

applied to the electrodes. These units are often referred to as rocker-arm machines. A major disadvantage of this design is that the upper electrode is moved through an arc. This makes it difficult to align the electrodes accurately, and uneven wear often occurs at the tip of the electrode, producing a misshapen weld.

Greater force and more accurate alignment can be achieved with hydraulic or pneumatic cylinders (fig. 12.10). These can be fitted to rocker-arm machines, but better results are achieved by direct coupling to the electrode, and most of the larger welders rely on this method of achieving pressure at the interface.

Fig. 12.10 *Larger spot welder with hydraulic pressure cylinder*

Portable units are widely used in the car industry. The transformer may be incorporated in the welding gun itself or be mounted remotely. Although the latter arrangement enables the gun to be lighter, it does mean that the cables to the electrode arms must be large, to avoid losses in the secondary circuit. Many of the smaller guns use a hand operated lever to apply force to the electrodes (fig. 12.11). Bigger units, and especially those for repetitive work, are fitted with pneumatic cylinders to allow higher pressures to be used and to reduce operator fatigue.

In a number of box-like fabrications, there are situations where it is possible to bring the electrodes into contact with only one side of the joint. Generally this means that conventional resistance spot welding cannot be used, but, where a copper bar can be placed against the other side, a technique known as *series welding* may enable acceptable welds to be produced (fig. 12.12). With this method, two electrodes are positioned about 50 to 150 mm apart and force is applied, pressing the sheets against the copper backing bar. Ideally, the current flows from one electrode, through the sheets, along the backing bar, and back through the sheets to the second electrode. There are thus two points on the interface where heating occurs and two spot welds are made simultaneously. In practical situations, some of the current flows along the top sheet, effectively reducing the amount of current flowing through the interface. Additional current must be supplied to compensate for this, and power requirements in series welding are therefore higher than in conventional spot welding.

189

Fig. 12.11 *Small portable gun with integral transformer*

Fig. 12.12 *Schematic diagram of series welding*

Resistance seam welding

The joints produced by spot welding cannot readily be made leak-tight, since they are bonded over only a small area of the interface. Admittedly, sealing compounds can be introduced between the overlapping sheets, but these are not always satisfactory, especially where flexing of the joint may occur, and their use increases the cost of the joining operation. A continuous bond or seam can be produced by overlapping the spot welds (fig. 12.13). After the first weld has been made, severe shunting occurs and the current must be raised to maintain the size of the welds. Accepting this limitation, satisfactory continuous seam welds can be made.

Fig. 12.13 *Overlapping spot welds used to produce a continuous bond at the interface*

In the simplest form of resistance seam welding, the operator moves the sheet by a set amount after making each weld. The technique is often called stitch welding. With more automated units, the electrode movement and weld times are controlled according to a predetermined cycle and the operator is concerned solely with moving the component between electrodes during the periods when they are open, i.e. when there is a gap between them.

A further development is to use disc electrodes (fig. 12.14), one of which is driven by an electric motor. As the electrodes rotate, the work is moved between them and pulses of current are supplied. Each pulse lasts long enough to produce a spot weld, and the time interval is arranged so that the spots overlap by about 30 to 40% of their length along the seam. Continuous movement of the electrodes may not be possible with metals that are difficult to weld, e.g. Nimonic alloys used in aero engines. In these cases the electrode wheels will be moved through a given angle and then stopped while the welding takes place. They are then indexed round to the next point and a further weld is made. This allows the weld cycle to be timed independently of the speed of rotation of the electrodes, thus giving more flexibility in choosing the optimum welding conditions.

Fig. 12.14 *Seam welding*

Projection welding

In both resistance spot and seam welding, current concentration or localisation at the interface is achieved by shaping the electrodes and restricting the area of the surface with which they are in contact. In projection welding, the weld is located by a projection raised on the surface of one of the sheets (fig. 12.15). When placed together, the sheets touch only at the point of the projection. Current flowing across the interface is concentrated through the projection, which heats up rapidly. As it becomes plastic and finally fuses, the projection collapses and a fused slug at the interface is formed. The result is similar to a spot weld.

Usually two or three projections are welded at the same time . The machines are essentially similar to those used for spot welding, but the electrodes are replaced by flat copper platens which exert a uniform pressure over the joint area. The choice of shape and size of projection is based on previous experience or experiment.

Projection welding is not often used for long lap joints. It finds much more useful applications in the joining of small attachments to sheet structures. It is widely used in the production of car bodies, domestic equipment, office furniture, and machine parts. For example, captive nuts can be made with small projections on one side which are used to attach them to a car chassis panel (fig. 12.16). Reinforcing rings are frequently projection-welded around holes in sheet-metal tanks. Threaded studs can be attached to backing bars by projection welding. A very important use of projection welding is in the manufacture of steel mesh for reinforcing concrete panels, making basket containers, and so on.

Fig. 12.15 Projection welding

Fig. 12.16 *Examples of projection-welded details*

12.3 Quality control in resistance welding

Resistance welding poses novel problems in controlling quality. NDT techniques such as radiography and ultrasonics yield very little useful information about spot, seam, or projection welds. It would also be impossible to justify the cost of testing all the spot welds in the majority of sheet fabrications. The alternative is to establish conditions which give an acceptable weld and to check that they are being maintained during a production run. A simple approach would be to make regular checks on the dimensions of the electrode tip to measure the amount of wear which effectively increases tip diameter and, hence, contact area resulting in lower current and pressure. At the same time, sample welds taken at random intervals could be tested. Some typical tests are illustrated in fig. 12.17. The drawback is, that to be completely effective, production should be temporarily halted to await the results of the test so that corrective action, if needed, can be taken before more defective welds are made.

A more sophisticated approach is to monitor the variables during welding. This is relatively straightforward for timing, current levels, voltage and pressure, but it is less easy to continuously measure tip wear. Also, factors such as sheet separation and surface condition (interface resistance) are difficult to assess. The interest in these variables, however, is the effect they have on the formation of the weld. The advent of microprocessor technology has opened up the possibility of monitoring the growth of the nugget, thereby giving a more direct measure of consistency of the finished weld. The information so gained during a production

193

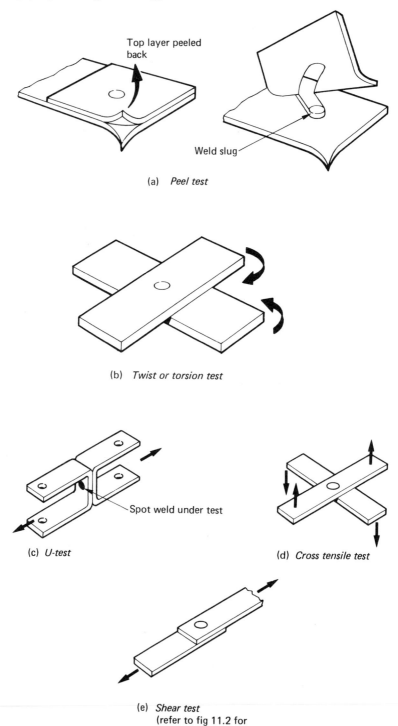

(a)   *Peel test*

(b)   *Twist or torsion test*

(c) *U-test*

(d)   *Cross tensile test*

(e)   *Shear test*
      (refer to fig 11.2 for
      mode of failure)

Fig. 12.17 Destructive tests on spot welds

run can be used to establish statistical control, but in the final analysis the ideal answer is to use error signals from the monitor to make on-line corrections using closed loop feedback control.

The extent to which these controls are exercised depends on the type of fabrication and the relative economics of the process. When welding flange joints in hot-air ducting, for example, it may be cheaper and just as acceptable to use simple equipment and increase the number of spot welds. On the other hand, the need for high quality in the welding of a flame tube in a jet engine justifies the inclusion of extensive quality-control measures.

12.4 Flash welding

The principles of resistance heating to obtain fusion are applied quite differently in flash welding.

The two parts to be welded must be of uniform cross-section. Normally, the process is used for joining bar, rod, or tubes, but there have been successful applications to sheet work. The parts are clamped with a small gap between the faces of the joint (fig. 12.18) and are connected to the output from a large

Fig. 12.18 Flash welding

transformer. There is a voltage of about 10 V across the gap so that a high current flows when the parts are moved into contact. As the surfaces are not smooth, contact is made at only a few points. The current flow is concentrated at these points, or bridges, which are rapidly melted by resistance heating. Immediately they are molten, the bridges are ruptured explosively, droplets of metal are ejected, and arcs are formed which heat the surfaces. There is now a gap between the parts. If the clamps are moved towards each other, contact is once again made and further bridges are established. The cycle repeats itself with more metal being ejected. This sequence is known as 'flashing' and is accompanied by showers of molten droplets over the surrounding areas. Flashing is an essential feature of the process, as during this period contaminants are removed with the molten droplets and the surfaces are heated to a uniform temperature. When these two tasks have been accomplished, a force is applied to the clamps, forging the parts together to give a solid-phase bond with any remaining molten metal being forced out to the surface. Here it forms a fin or flash around the joint. For some applications, such as pipe welding, it may be necessary to remove the flash, as it could obstruct the bore.

The equipment used for flash welding tends to be bulky. The clamps must exert sufficient pressure to prevent the components from slipping when the force is applied to bring them together. They must also restrain lateral movement, as correct alignment of the parts is essential if a uniform distribution of heat is to be achieved. At the same time, the clamps must be mounted on a rigid bed which does not flex or distort during the welding sequence.

Flash welding makes appreciable demands on the electrical power supply to a welding shop. Currents in excess of 100 000 A can flow across the interface, and power inputs of up to 200 kVA may be required. As the transformers used in flash welding are single-phase, an unbalanced load is placed on the 440 V three-phase supply used in most factories in the UK. Frequently this leads to problems which can be solved only by the installation of special transformers designed to distribute the load uniformly over the three phases.

Nevertheless, flash welding offers many attractions in the volume production of square or tubular components in a range of materials. It is particularly useful for the joining of non-ferrous metals such as Nimonic alloys and titanium. The absence of molten metal in the completed joints means that weld-metal cracking problems often associated with fusion welds are avoided, and that dissimilar metals can be joined. Some typical applications are the butt welding of continuous railway track and of boiler tubes, the manufacture of chain links, and the fabrication of metal window frames.

Usually, the joint is in the form of a butt with the axes of the components in line, but corner joints between sections and some T joints are practical propositions. The main limitations are the maximum force which can be applied (the average requirement is about 70 N/mm^2 for low-carbon steel and up to four times this for heat-resistant metals such as Nimonic alloys) and the cross-sectional area which can be accommodated by the power supply available .

13 Welding with power beams

In arc fusion welding, the width of the weld pool at the surface is usually between 5 and 10 mm, which means that the heat from the arc is spread over an area ranging from 40 to 180 mm^2 depending on the voltage, current, and travel speed. As the heat flows into the joint, the weld pool assumes its characteristic saucer-shaped profile. A radically different approach to weld formation is used in a relatively modern group of three processes: arc-plasma, electron beam, and laser welding. In each of these, the fusion and bonding are achieved by a technique known as 'keyholing', involving the use of a high energy beam.

13.1 Keyholing

We can see how keyholing works by considering a beam directed at the surface of a metal plate (fig. 13.1). The beam has a small diameter so that it can be

High-energy beam

Energy from beam melts surface.

Molten metal pushed to the side to form crater.

Beam melts more parent metal. A new crater is formed.

Sequence is repeated until beam penetrates through thickness.

Metal condenses and solidifies at trailing edge, giving a weld between the components.

Beam

Beam energy melts metal at leading edge through thickness.

Fig. 13.1 Principles of keyholing

focused on to a spot on the plate surface about 1 or 2 mm in diameter. It also has a high kinetic energy, which it releases when it collides with a solid object. The amount of energy released needs to be in excess of 10 kW/mm2 if a satisfactory weld is to be made.

At the point of impact, the temperature of the metal rises rapidly over a small area. The metal melts, and some vaporises. The molten metal is forced to one side, forming a crater exposing solid metal at the bottom of the small pool which has been formed. The beam can now impinge on the solid material and in doing so releases further energy. Melting of this newly uncovered metal occurs, another crater is formed, and the cycle is repeated until the beam has penetrated through the plate. At this stage there is a cylindrical cavity or *keyhole* extending through the thickness of the plate. The wall of this hole is covered by the molten metal which has been forced outwards from the centre-line of the beam. The metal is held in place by surface tension and by the pressure of metal vapour present in the hole.

We now have a situation in which the beam is travelling through the hole and emerging from the reverse side of the plate with little reduction in energy. If, however, we move the beam to one side, it comes into contact with the wall of the hole and releases its energy. The hole becomes temporarily elongated, heat is lost from the area vacated by the beam, and some of the metal on the trailing wall of the hole solidifies. Metal melted from the leading edge flows under surface tension round the hole to restore its circular cross section. As the beam continues to move across the plate, this process of melting and solidification proceeds at a regular pace, keeping the shape of the hole uniform. At the completion of the traverse, a narrow band of cast metal has been produced through the thickness of the plate along the line of travel of the beam.

The sequence described is called keyholing. It offers a useful technique for welding since, by moving the beam along the line of a square edge joint which has not more than a small gap between the mating faces, the molten metal coating the keyhole wall bridges the interface. Progressive solidification of the molten metal on the trailing wall bonds the two components of the joint, as in other fusion-welding systems.

Welds made by keyholing are characterised by their large depth-to-width ratio, which can be as high as 20:1 (fig. 13.2). In conventional arc fusion welding, where solidification proceeds from the fusion boundary to the centre, giving large grains, there would almost certainly be cracks in welds with such high depth-to-width ratios, assuming that they could be produced. Successful results are possible with keyholing because only thin layers of molten metal are involved in the solidification process at any given time. Small grains are formed, and the weld metal has good strength at high temperatures. The problem of ensuring an adequate supply of liquid metal to accommodate shrinkage is also made easier, since solidification proceeds along the line of travel, not vertically through the weld thickness.

Keyholing offers two major advantages: uniform-penetration beads can readily be obtained and the parallel-sided weld gives uniform shrinkage with little or no distortion. Although a number of possible energy sources have been proposed for key holing, only three systems are currently in use: plasma-arc, electron-beam, and laser.

198

Fig. 13.2 Cross-section of electron-beam weld (scale graduated in cm and 2mm divisions)

13.2 Plasma-arc systems

The term plasma means a body of ionised gas. Plasma is produced when a gas is heated to a temperature which is high enough to dissociate it into positive ions and negatively charged electrons. Energy is required to bring about this dissociation, and in welding this energy is supplied by the arc. The gas in the centre or column of the arc is dissociated at the temperatures involved, creating a plasma. As this gas flows away from the arc column, it reassociates to produce neutral atoms, giving up its energy in the form of heat.

The temperatures experienced in a tungsten arc used for TIG welding are of the order of 11 000°C. The arc column is bell-shaped (fig. 13.3) and is free to

Fig. 13.3 Plasma-arc welding

move within the gas shield. If the arc is surrounded by a water-cooled copper tube, it is constrained and the temperature in the arc column increases to about 20 000°C. If the plasma gas is forced to travel through this constricted arc, it is heated to temperatures much higher than those in the conventional welding arc. It expands rapidly and issues from the constricting orifice as a high-temperature ionised gas jet. In other words, it is the type of high-energy beam required for keyholing.

Using this, butt welds can be made in material of 3 to 15 mm thickness in one pass with currents of 150 A to 300 A. Speeds are generally between 40 and 80% higher than for TIG welding of the same joints. The process is not widely used, as its range is limited, but successful pipe-welding techniques have been developed to take advantage of the very good penetration bead profiles that are characteristic of keyholing.

A low-current version, known as micro-plasma welding, has proved to be more successful. The system is operated at 0.1 to 10 A and uses nozzles much simpler than those required for plasma-arc welding. Micro-plasma welding is of particular value in the welding of metal less than 1 mm thick, as it offers stable operation at low currents – this is difficult to achieve with TIG welding (page 93). The welds are small and are more like those found in conventional arc welding. It seems unlikely that keyholing can operate effectively on very thin material when using plasma jets.

13.3 Electron-beam welding

An electron-beam welding machine has three main components (fig. 13.4):

a) a gun which produces a controlled beam of electrons,
b) a vacuum chamber with associated pumping equipment,
c) a unit which either traverses the beam along the joint line or moves the work under the gun.

Fig. 13.4 *Main components of an electron-beam-welding machine*

Electron gun

Various types of gun have been used in commercial equipment for electron-beam welding, but in the main they are based on the same principles as the cathode-ray tubes and thermionic valves found in television sets. The source of electrons is a heated tungsten filament wire mounted in a cup shaped electrode, which helps to control the flow of electrons (fig. 13.5).

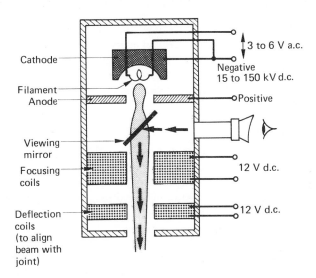

Fig. 13.5 Main components of an electron-beam gun used for welding

The anode is usually in the form of a disc with a hole in the centre. Two ranges of voltage are used. In the low-voltage systems, the difference between the anode and the cathode is typically 15 to 30 kV, while a range from 70 to 150 kV is common for high-voltage machines. A number of operating advantages are claimed for both systems, but the resultant welds are essentially similar. At the voltages used in these guns, the electrons move rapidly from the cathode to the anode, where some are collected. Others pass through the hole in the centre of the anode and continue to travel until they meet a solid object, when they give up their kinetic energy in the form of heat.

The electrons which emerge from the hole in the anode are not necessarily travelling in the same direction, and the beam is usually divergent – in other words, it tries to take the form of a cone with the apex at the anode and the base on the surface of the work. If this were allowed to happen, the electrons would be spread over a relatively large area of the surface and it would not be possible to attain the high energy intensity which is essential for keyholing . The gun therefore contains magnetic coils which deflect the electrons towards the central longitudinal axis of the beam. By controlling the current flowing in the coils, and hence the strength of the magnetic field, the beam can in this way be focused to a diameter of about 0.5 mm.

With such a small diameter, it is of the utmost importance that the beam should be accurately aligned with the joint. Viewing windows can be fitted into the walls of the vacuum chamber. A much more satisfactory method is to sight along the axis of the beam by using a telescopic eyepiece and angled mirrors.

Vacuum chamber

If electrons are projected through air, they lose their energy in collision with oxygen and nitrogen atoms. Hence it is essential to operate in a vacuum at a pressure of 10^{-4} to 10^{-2} torr (0.013 to 1.3 N/m^2) if the beam is to travel any distance.

The chamber must be large enough to accommodate the work and the gun. It must also be designed to protect personnel against radiation hazards. When the electron beam collides with the surface of the work, some of the energy released is converted to X-rays. These must not be allowed to escape, as they can harm workers in the vicinity, and they are contained by applying a lead lining to the chamber.

Traversing the joint

If the joint is small, it may be possible to complete the weld by moving the beam rather than the work. This has the advantage that mechanical traverse systems are not required – the beam can be deflected from its central line by magnetic coils. Current flowing in the coils is programmed to direct the beam along a predetermined path. The maximum deflection is relatively small, however, and for the majority of work the gun and the component must be moved relative to each other. In general, it is easier to move the work using a motor mounted outside the chamber. Because the traversing gear operates in a vacuum, dry lubricants must be used, as normal greases and oils evaporate and contaminate the chamber.

Applications

Electron beam welding is particularly useful for the joining of highly reactive metals such as titanium and zirconium. These are kept free of contamination by the vacuum. The process is not limited to these materials, however, and many applications make use of the other unique attributes of this particular method of keyholing:

a) deep penetration, enabling one-pass butt welds to be made in plate up to 75 mm thick;

b) virtually uniform shrinkage about the neutral axis of the plate and narrow heat-affected zones, leading to very low levels of distortion – this allows fabrication of components without the need for post-weld machining;

c) removal of gases from the weld metal by virtue of the vacuum – for example, hydrogen is reduced to levels at which successful welding of hardenable steels can be achieved without the formation of heat-affected zone cracks (see page 109).

The high cost of vacuum chambers tends to limit the use of electron-beam welding to moderately sized components. In industrial practice up till now it has not been a major competitor to conventional fusion welding but has opened up new possibilities, especially in the manufacture of machinery. Gear clusters for transmission systems, intricate valve arrangements made from corrosion-resist-ant alloys, and pressure capsules are a few examples of components which can be welded in the pre-machined condition without significant loss of dimensional tolerances. Such items can be used in the as-welded condition without the need for the post-weld heat-treatment and machining operations often associated with arc fusion welding of similar work.

The component shown in fig. 13.6 is an example of a design which it would be difficult and expensive to machine from the solid. It can be readily made by machining three separate parts which are then electron-beam-welded. Only a small amount of machining is needed to remove the excess metal from the outer surfaces of the welds before the component is ready for service.

= Electron beam weld

Fig. 13.6 *Example of electron-beam-welded machined pump component*

The electronics industry has made extensive use of electron-beam welding. The ability to locate small welds accurately is a considerable advantage in sealing capsules. Perhaps of more importance in many applications is the protection afforded to the materials used – welding in a vacuum means that there is no risk of oxidation which could impair the operation of semiconductors.

At the other end of the scale, advances in chamber and gun design have enabled large components in thick materials to be welded. The chamber illustrated in fig. 13.7 has a useful working volume of 7 m × 3.6 m and is fitted with a 75 kW gun. It can be pumped down to an operating pressure of 5×10^{-2} torr (7 N/m^2) in about 35 minutes and it can weld steel in thicknesses up to 150 mm. A 3 m long butt weld in this thickness would take about 20 minutes to complete.

13.4 Laser systems

The heating effect of concentrated rays of light is well known. Sun rays focused on to a spot by a magnifying glass can make paper burn. By directing all the rays to one spot, the energy density is increased and the paper is raised to ignition temperature.

The amount of energy that can be harnessed in this way is limited. It is relatively easy to ignite a piece of paper, but the effect on a piece of steel would

Fig 13.7 *Large electron-beam-welding chamber*

be negligible. Heat would quickly be conducted away, and the temperature would rise only a few degrees.

Visible light consists of a number of radiations of different wavelengths. The waves travel in random directions and are not in phase. Before we can achieve the high energy densities needed to melt metal, the radiations must be converted to the same wavelength and brought into phase using a laser.

The name 'laser' is an abbreviation of Light Amplification by Stimulated Emission of Radiation. In a simple laser (fig. 13.8), flashes or pulses of white light are directed into a YAG crystal. (YAG is an abbreviation of yttrium aluminium garnet.) The YAG crystal absorbs the energy and converts it into a single-wavelength beam of infra-red light with a diameter of a few millimetres. The output of a YAG laser is limited by its inability to operate at high average powers. Current commercial equipment is limited to about 400 W average power, giving weld penetrations of about 1 mm. For deeper penetration welding we need continuous operation at higher levels of power (up to 15 kW).

In commercial higher-power lasers used for welding, the YAG crystal is replaced by a tube filled with a mixture of carbon dioxide, nitrogen, and helium gases (fig. 13.9). A high voltage developed across electrodes placed inside the

Elliptical tube with reflective inner surface which focuses light rays on to YAG rod.

Discharge tube generates pulses of white light.

Laser beam emerges from end of YAG rod.

YAG rod (crystal) collects light rays.

Fig. 13.8 Principles of a YAG laser

tube makes the gas fluoresce. The infra-red rays amplified within the discharge are reflected by mirrors at the ends of the tube, and the laser beam which emerges is focused to a point on the work. Output powers up to 20 kW are obtainable with CO_2 lasers.

Laser welding

In many ways, laser welding is similar to electron-beam welding. As the beam hits the surface of the work, kinetic energy is released and a weld pool is formed. At lower power levels, the weld pool is saucer-shaped and is similar to that of an arc weld. Use is made of low-powered lasers in the electronics industry for spot welding – typically for joining wires to flat surfaces and for securing edge joints. With power levels above 1.5 kW, keyholing can take place and the welds take on the characteristic profile which, as we have seen, is associated with electron-beam welding (fig. 13.10). Welds of this type can be made in thicknesses up to 10

Mirror reflects infra-red rays produced by discharge.

Pumping unit and heat exchanger to cool gas

Discharge

Mirror with window to allow coherent infra-red radiation to escape

Mirror

Laser tube filled with carbon dioxide/ helium/nitrogen mixture

High voltage causes fluorescent discharge in tube.

Shielding gas

Work

Fig. 13.9 Outline diagram of a CO_2 laser

Fig. 13.10 *Profile of a typical laser weld in 10 mm steel*

mm with commercially available lasers; but, above 6 mm, travel speeds are down to less than 1 m/min and stability problems in the weld pool can arise at these slow rates. One way of offsetting this is to use a wire-feed attachment to make a multipass weld, adding filler metal for the second pass (fig. 13.11).

Unlike electron-beam welding, however, there is no need for the vacuum chamber with laser welding, as the laser beam travels readily through air. While this is a considerable advantage, it does mean that the weld is exposed to

Fig. 13.11(a) *Wire-feed attachment for laser welding*

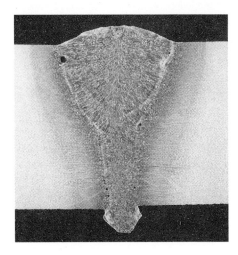

Fig. 13.11(b) *Macrosection of a two-pass laser weld*

atmospheric contamination. The molten metal must be protected by a helium or argon gas shield in a similar way to TIG welding.

The application of laser welding to industrial practice is well established. The electronics industry makes wide use of the technique, and there have been examples of successful laser welding of components for nuclear plant. More recently, manufacturers of cars and other domestic products have been incorporating lasers into their production lines. In the car industry the process has been used to weld transmission components in competition with electron beam welding; the absence of a vacuum chamber in laser welding reduces down-time. Smaller units such as fuel injectors and filters are also laser welded.

Like electron-beam welding, a major attraction of laser techniques is the limitation of the heating effects. This is well illustrated in the manufacture of heart pacemakers. The electronic unit and its battery are encapsulated in a metal box or can. Because the battery and the electronic components would be damaged if they were raised to too high a temperature, soldered cans have to be made larger than would otherwise be necessary, to keep the components away from the heated area. The relatively small spread of heat in laser welding has meant that the size of the containers can be reduced appreciably, to the benefit of the patient.

14 Thermal-cutting systems

A number of thermal-cutting systems have been developed in which metal is melted along the cutting line and is blown out of the cut by a gas jet. Some of these have been widely used while others have found restricted application.

In fabrication, the most commonly used techniques for thermal cutting are:

- oxygen cutting
- oxygen-arc
- air-arc
- plasma-arc
- laser.

14.I Oxygen cutting

Undoubtedly the most frequently used thermal process for the preparation of steel plate is oxygen cutting.

To produce a cut, the steel is first heated by an oxygen-fuel-gas flame until it has reached the temperature at which the iron ignites when it is brought into contact with an atmosphere of high-purity oxygen. This is known as the oxygen-ignition temperature and is about 950°C. A jet of pure oxygen is directed at the heated area (fig. 14.I), 'burning' the steel in its path. Heat is released which melts

Fig. 14.I(a) Principles of oxygen cutting

Fig. 14.I(b) *Oxygen cutting of steel plate*

surrounding metal and pierces a hole in the plate. If the nozzle is moved along the cutting line, so that the steel is progressively heated by the preheat flame and then oxidised by the oxygen jet, a cut is produced. The gap or kerf created by the cutting jet can be kept constant to within ± 0.5 mm on plates up to 50 mm thick.

A cut of high quality can be obtained if the correct conditions are used and the speed of travel of the cutting jet is uniform. The latter is difficult to achieve if the blowpipe is hand-held, although mechanical aids are available. In general, machine cutting in which the nozzle is traversed mechanically along the cutting line gives a higher quality of cut.

Typical travel speeds for oxy-acetylene cutting are given in Table 14.I.

The chemistry of oxygen cutting

The key reactions in oxygen cutting are those between iron and oxygen since, not only do they convert the steel to an oxide slag for easy removal from the kerf, they

Table 14.I *Typical travel speeds for oxygen cutting*

Plate thickness (mm)	Travel speed (mm/min)	
	Square edge	Bevel edge
6	420	370
12	360	320
25	250	220
50	180	170
100	160	150

209

also provide heat to keep the reaction going along the length of the cut. The two important reactions are:

$$3Fe + 2O_2 \longrightarrow Fe_3O_4 + 1100kJ$$

$$Fe + O \longrightarrow FeO + 270kJ$$

As the cutting jet meets the hot steel at the top of the cut, some of the oxygen reacts with the metal. Heat is released, and a molten iron-oxide slag is produced. The heat also melts some of the metal beneath the cutting face, and molten steel is mixed with the iron oxide. Slag analyses show that about 30% of the iron in the material ejected from the bottom of the cut is not oxidised and hence was removed as molten steel.

The slag produced by these reactions is blown down the cut by the jet. Heat in the slag raises the temperature of the next part of the cut, which is then brought into contact with the oxygen as the slag continues to be blown away. Further oxidation of the slag occurs, and more heat is generated. In this way the reaction proceeds through the thickness of the metal. Heat liberated at the top is carried down the cut, ensuring that the metal is always at the oxygen-ignition temperature when it is progressively exposed to the cutting jet. The only area which is not heated in this way is the top surface of the plate, since the heat generated there is carried down into the kerf. We must therefore use the preheat flame to keep the surface of the plate above the ignition temperature, otherwise the reactions do not take place and cutting ceases.

Formation of a boundary layer

If the sequence of events we have just described is to proceed at a regular rate, it is imperative that the cutting oxygen must come into contact with 'clean' hot metal. In other words, there must not be any barrier which could prevent the oxygen from freely reaching the steel and thus slow down or even stop the reactions. How could such a barrier arise? So far we have only discussed reactions between iron and oxygen, but there are other alloying elements in steel which oxidise to form gases. Carbon produces a mixture of carbon monoxide and carbon dioxide, and sulphur is oxidised to sulphur dioxide. These gases are generated at the interface between the cutting oxygen and the steel, and must be swept away by the velocity of the oxygen jet. If they are allowed to collect, they form a stagnant boundary layer through which the oxygen diffuses only slowly or not at all (fig. 14.2). The amount of oxygen available at the surface of the steel is thus reduced, and insufficient heat is generated to keep the metal above the ignition temperature. The velocity of the cutting oxygen jet is therefore a critical factor in achieving cuts of acceptable quality.

Metallic oxides arising from the oxidation of alloying elements in the steel will not interfere with the cutting operation unless they increase the melting point or viscosity of the slag. A metal which behaves in this way is chromium, which is added as an essential element in stainless steel to obtain resistance to corrosion. The chromium oxide formed at the oxygen-metal interface raises the melting point of the slag and increases its viscosity. The slag is not readily removed from the reaction area, and an impervious boundary layer is established which prevents satisfactory cutting. The only practical solution to this problem is to add flux via the oxygen stream to reduce the melting point and viscosity of the slag.

Fig. 14.2 *Formation of boundary layer in oxygen cutting*

The quality of the cuts produced in this way is not of the same standard as that achieved on plain-carbon steels.

The preheat flame

The preheat flame has two main functions: firstly, it raises the starting point to ignition temperature; secondly, it keeps the top layers of the steel hot so that the iron-oxygen reaction can begin as soon as the cutting jet impinges on the surface.

In theory, the choice of fuel gas is not critical, since any gas which when mixed with oxygen keeps the surface at the ignition temperature is acceptable; in practice, the choice of gas for the preheat flames plays an important role in the economics of the process. The gases most commonly used are acetylene, propane, butane, and proprietary gases based on hydrocarbon compounds. The latter are sold under a variety of trade names which give little indication of their composition. Some use propylene which has the chemical formula $CH_2:CH.CH_3$. Others are based on a more complex compound: methylacetylene-propadiene stabilised (abbreviated to MAPS).

The choice of the most suitable gas depends on a number of factors, and each application must be evaluated separately. Amongst the factors to be taken into account are gas cost, cutting speed, time to start a cut, ease of handling and supply to the cutting blowpipe, and availability of suitable nozzles.

The cost of the fuel gas by itself is of little value as the amount of oxygen needed to achieve a neutral flame varies from one fuel gas to another. Typical ratios are:

oxygen:acetylene	1.5:1
oxygen:MAPS	2.5:1
oxygen:propylene	2.6:1
oxygen:propane	4.5:1

Acetylene gives the highest flame temperature and it requires the least time to raise the metal at the starting point to ignition temperature. Propylene and MAPS take slightly longer whilst propane is significantly longer still. Where the starting time is an appreciable proportion of the total cutting time, acetylene can be a very attractive choice. On long cuts, on the other hand, start time becomes less important and other factors may lead to the choice of one of the other gases.

Storage and handling is an important consideration in choice of gas. With the exception of acetylene, the fuel gases can be liquified and stored in cylinders. Acetylene can be compressed to a liquid, but it is liable to be unstable when in this form and can explode. It is stored, therefore, in cylinders which contain a porous material soaked in acetone. Under pressure, acetylene dissolves in the acetone and is in a safe condition for use provided the cylinders are kept in a vertical position with the valve uppermost. Where an appreciable amount of cutting is to be done, liquified gases can be stored in tanks and distributed to the working points, thus avoiding the need for large numbers of cylinders. Acetylene, however, cannot be handled in this way. Where the demand for acetylene is high, a number of cylinders can be connected together by a manifold.

A secondary function of the preheat flame is to burn off contaminants such as paint and grease from the surface of the plate ahead of the cutting stream. It can only cope with small amounts, however, and badly contaminated surfaces must be cleaned before the cutting operation is started.

Nozzles for oxygen cutting

Amongst the factors mentioned in the discussion of fuel gases was the need to have a suitable nozzle. Nozzles for cutting are machined from copper-alloy and have a hole drilled through the centre for the cutting oxygen (fig. 14.3). This is surrounded by ports or holes through which the mixed fuel gas and oxygen flow for the preheat flame. The diameter of the cutting oxygen orifice, which determines the flow rate of the gas, is chosen to match the thickness of steel to be cut and is independent of the type of fuel gas. On the other hand, the diameters of the preheat ports must match the volume of mixed preheat gases which depends on the fuel gas being used. The amount of oxygen needed to burn propane is

Fig. 14.3 *Nozzles for oxygen cutting*

significantly greater than for acetylene and larger preheat ports are required for effective operation.

Although some blowpipes supply mixed fuel gas and oxygen to the nozzle, it is more usual for these gases to be mixed in the nozzle itself. Nozzles for use with acetylene are made from one piece of copper alloy (fig.14.3). This design, with suitable preheat port sizes, can be used for the other gases but more commonly the nozzle is made from two pieces. The central part of a two-piece nozzle is fluted to form preheat ports at the outlet.

Effect of oxygen cutting on parent-metal structure

Since gas cutting is basically a thermal process, a heat-affected zone is produced in the parent plate adjacent to the cut edge. In steels with very low carbon contents this is of little significance, since at the cooling rates experienced the steel is unlikely to harden. With higher carbon contents, bainitic and martensitic structures may be produced in the steel and the cut surface becomes hard. Provided this hardened zone does not contain cracks, the edge is acceptable for most welding applications. The extent of the heat-affected zone (fig. 14.4) depends principally on the plate thickness but also to some extent on the type of cut, i.e. square or bevel. Where the hardening effect is considered detrimental, heat treatment can be applied to temper the cut edge.

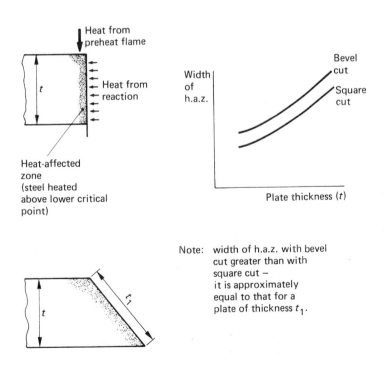

Fig. 14.4 Heat-affected zone in oxygen cut edges

213

Equipment for oxygen cutting

The equipment used for gas cutting is similar to that described for oxy-acetylene welding. A supply of fuel gas and oxygen is required, and either cylinder or bulk storage (fig. 14.5) can be used. Since large quantities of oxygen may be needed during the cutting operation, it is more common in large fabricators' works to find bulk-storage tanks. Independent control of the preheat and cutting-oxygen supplies is essential.

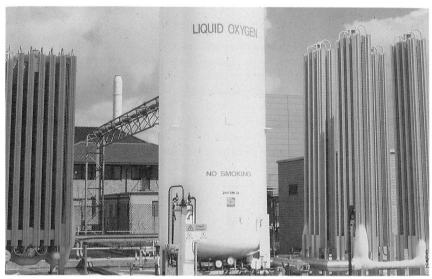

Fig. 14.5 Bulk storage tank

One of the major problems which used to be encountered in oxygen cutting was backfiring in the preheat flame, associated with burn backs. The use of non-return valves and flash-back arresters contributes to the safety of the operation, but the real answer lies in the careful design of torch and nozzle. As long as they are properly maintained and the correct pressures are used, modern torches give uniform burning of the oxygen-fuel-gas mixture at the tip of the nozzle without the risk of backfiring. Badly worn nozzles or damaged nozzle seatings create situations under which this problem can recur.

The nozzle and torch systems used for manual and mechanised gas cutting are essentially similar – the main difference between the two techniques lies in the better control of nozzle-to-plate distance and traverse speed which can be achieved by mechanisation. The simplest machines use a motorised cross slide to move the cutting head across the plate. Slightly more complex units contain a facility for movement in two directions. Coupled with a template, this provides a method of profile cutting (fig. 14.6). Large machines can have more than one cutting head, and each head can be fitted with two or three nozzles to cut double bevel-edge preparations, with or without a root face, in one operation. These heads are motorised and can be guided by a variety of methods – templates, photoelectric followers, and numerical-control systems are probably the most popular techniques in current practice.

214

Fig. 14.6 Typical profile cutting machine

14.2 Oxygen-arc cutting

The principles of oxygen cutting can be applied to arc cutting. Since the preheat flame in oxygen cutting is principally used to heat the metal to the ignition temperature and then maintain it at this level, it can be replaced by other methods of heating, e.g. an arc.

In oxygen-arc cutting, a hollow electrode which has a flux covering is used. The electrode is connected to either a d.c. or an a.c. power-supply unit, although d.c. with electrode negative is usually preferred since it tends to give faster cutting speeds. The arc is struck at the edge of the plate, and the metal is rapidly heated to the ignition temperature. Oxygen is supplied to the hole in the electrode at a pressure of about 5 bars. The cutting jet thus produced impinges on the hot metal and an iron-oxygen reaction takes place as in oxygen cutting (fig. 14.7). The electrode is moved along the plate, with the outside edge of the flux covering in contact with the surface. Since the end of the electrode burns in the form of a cone, the arc length is thus kept constant. The travel speed is adjusted to give a uniform kerf width.

The flux covering assists in cutting steels containing alloying elements which would otherwise retard or stop the exothermic reaction between iron and oxygen. Although cuts can be achieved on oxidation-resistant metals, the mechanism is quite different, since the arc provides the heat for melting and the oxygen jet is used to blow the molten metal out of the kerf.

Two sizes of electrode are commonly used: 5 and 7 mm diameter with holes which are 1.6 and 2.5 mm diameter respectively. The electrodes are usually 450 mm long, and the length that can be cut with one electrode depends on the thickness of the material.

Fig. 14.7 *Principles of oxygen-arc cutting*

Data for oxygen-arc cutting of low-carbon steel

Table 14.2 gives typical data for oxygen-arc cutting of low-carbon steel; cutting speeds for those metals where the heat is supplied principally by the arc and not by an exothermic reaction are appreciably slower. For example, the speed for 25 mm stainless steel or Monel plate would be 4 m/h, while with bronze of the same thickness the figure would be 5 m/h compared with about 30 m/h for low-carbon steel.

In general, the surface finish of the cut is rougher than that achieved with oxygen cutting or arc-plasma cutting. The process can be used in a mechanised form, but most applications involve manual operation. Guide bars can be readily

Table 14.2 *Data for oxygen-arc cutting of low-carbon steel*

Electrode size (mm)	Thickness of metal (mm)	Current (amperes)	Cutting speed (m/h)	Length of cut per electrode metre
5	8	120	57	1.14
5	19	125	38	0.74
5	38	135	26	0.58
7	50	190	26	0.61
7	75	230	17	0.42

used, since the insulating properties of the flux covering prevent shorting out when the electrode touches the bar. The depth to which the heat penetrates is smaller than in flame cutting, and higher cooling rates can occur, with a greater attendant risk of hardening in carbon and alloy steels.

14.3 Air-arc cutting and gouging

The air-arc technique for cutting is significantly different from oxygen or oxygen-arc cutting. All the heat is provided by an arc, and a jet of compressed air is used to blow the metal out of the cut (fig. 14.8).

The electrode material in air-arc cutting is a combination of carbon and graphite, and usually it is coated with copper. The carbon electrode is held in a specially designed holder containing holes through which jets of compressed air blow along the electrode. The electrode is connected to the positive terminal of a d.c. supply unit, and the arc is struck between the electrode and the surface of the parent metal. A molten pool is established, and the metal is blown from beneath the arc by the compressed air. Further melting takes place followed by removal of the metal until the cut has penetrated the thickness. The arc is then traversed along the line of the cut. The surface finish is not good compared with that produced by flame cutting, but, since the heat is supplied by the arc, both non-ferrous and ferrous metals can be cut.

Compressed air

Cable to power-supply unit

Air jet

Copper-coated carbon electrode

Molten metal and slag blown from cut

Fig. 14.8 Principles of air-arc cutting

217

The air-arc system has only a limited application in cutting. More commonly it is used as a gouging tool to prepare the reverse side of a weld for the deposition of a sealing run or to remove metal which contains defects. In the hands of a skilled operator, grooves can be produced which require little preparation before being filled with weld metal.

14.4 Plasma-arc cutting

The principles underlying the plasma-arc welding system described in Chapter 13 (page 199) can be applied to cutting by increasing the velocity of the plasma jet. The keyholing mode of operation is retained but metal is blown through the hole, thus preventing the formation of a 'bridge' after the arc has been moved forward.

The general configuration of the torch is similar to that used for welding. The gas surrounding the arc can be argon, but better results are obtained by using a gas which increases the arc voltage. Argon-helium and argon-nitrogen mixtures give good operating conditions; nitrogen can also be used alone. The higher arc voltage coupled with a faster gas flow means that a high open circuit voltage is required. This is normally at least 120 V, and safety devices must be built into the equipment to avoid possible contact between the operator and the electrode. An optional outer gas shield of air or nitrogen helps to cool the torch; it is also claimed that the quality of the cut is improved. The plasma jet, travelling at supersonic speeds, generates high levels of noise, making the wearing of ear protectors essential. The cutting operation can be operated under water, however, significantly reducing the noise level; in this case, the secondary shield is not required.

Plasma arc cutting relies on a mechanical action to remove metal from the kerf, and this technique does not involve a chemical reaction. It can be used, therefore, on a range of non-ferrous metals which are not suited to oxygen cutting. Accurate, high quality cuts can be made in aluminium (up to 150 mm thickness) and stainless steel (up to 100 mm). Cutting speeds are very fast. With aluminium they range from about 100 mm/min on 150 mm thickness to 7000 mm/min at 6 mm. The process can be used economically on carbon steels in thicknesses below 13 mm where the savings made by high cutting speeds can be used to offset the expense of the plasma equipment. Above this thickness, oxygen cutting is the normal choice.

The high cutting speeds which are available with plasma arc cutting mean that attention has been focused on mechanised applications. Manual cutting is possible, but the constraint placed on the operating variables to reduce the cutting speed to manageable levels inevitably results in a reduction in the quality and accuracy of the cut.

14.5 Laser cutting

One of the first applications of laser techniques to metals was for cutting, and it still accounts for the widest use of high power lasers in the fabrication and engineering industries.

As with plasma-arc cutting, laser cutting relies on keyholing to penetrate through the thickness, and then it blows molten metal out of the hole by a gas jet. A nozzle is fitted concentric with the output from a laser so that the jet can be directed at the metal surface coaxial with the laser beam. The jet can be an inert gas, nitrogen or, in the case of steel, oxygen. In the last of these, the oxygen reacts exothermically with the steel, generating additional heat.

Laser cutting can be used on carbon steel and stainless steel in thicknesses up to about 10 mm, as well as on a range of non-ferrous metals, but in thinner sections only. It is not confined to metals. Wood, ceramics, plastics, cardboard and synthetic clothing materials can all be cut by laser beams, often without the need of a gas jet.

15 Which process?

15.1 The range of factors influencing choice

In the preceding chapters we have discussed many different welding processes. Each has its own particular characteristics and attractions. Some are widely used, others are known only in specialist circles. With so many alternatives on offer, what are the factors governing the choice of any one of them? This raises a series of questions:

1) Why do so many fabricators choose MMA to weld steel plate?
2) What are the attributes of TIG welding which make it more attractive than MMA to the aero-engine industry?
3) Why is resistance spot welding almost always the first choice of the car-body manufacturer?

How in practice do we decide which process to use for any particular application? It would be easy to say that the process chosen should provide the required quality at the lowest cost. While this must always be our aim, there are usually constraints which make our decision a compromise. We have earlier discussed the problems encountered in trying to define quality in welded joints. In making a choice between processes, we must consider quality in terms of the skill of the welders available to do the job. Similarly, availability of equipment contributes an important restraint – the volume of work may not justify investment in new plant or retraining labour, and existing welding sets are often used even though the apparent cost may be higher.

A further problem arises in determining the true cost of a process. At first sight it should not be difficult to work out the costs of a welding operation. Consumables and welding times can be readily measured. From the data obtained, a price per metre of welded joint can be deduced for each process and the results can be compared. But this does not give the complete picture. We must take into account the interaction between welding and other manufacturing operations. Some edge preparations cost more to cut than others. Welding processes and procedures which give significant amounts of distortion involve increased finishing costs in the correction of shape by straightening, and so on. Where the surface of the finished joint must be flat, the use of a technique which gives a smooth profile with a minimum of excess metal and spatter leads to savings in post-weld grinding or machining. In the fabrication of small machined components, a welding process which does not produce fume, spatter, etc. and which can be used alongside plant such as lathes and shapers offers marked savings in movement costs and production time. These few examples illustrate the need to consider welding as an integral part of an overall manufacturing operation. They also indicate some of the difficulties met in attempting to produce a true cost for welding.

The restraints we have identified vary from one situation to another. While recognising their existence, if we are to make sensible comparisons in general terms, we must have freedom of choice. In the following pages we need to assume that equipment and skilled labour are readily available and that ancillary fabricating operations such as cutting can be adjusted to suit each welding process.

15.2 Technical considerations

The first task is to ensure that the process chosen satisfies the technical requirements of the operation. The aspects which need to be considered are conveniently grouped under three headings:

- type of material,
- type of joint, and
- production requirements.

A variety of questions can be asked about each of these aspects. The following list is not exhaustive – there are many more that could be asked in evaluating an application. The questions presented below aim to indicate the various factors which need to be considered.

Type of material

a) Will the process produce satisfactory, controlled fusion of the joint faces?
b) Can the weld metal be adequately protected against atmospheric contamination? As an example we can compare low carbon steel with zirconium. With the former, all processes are satisfactory. Zirconium, however, is a very reactive metal and rapidly absorbs oxygen and nitrogen. Acceptable welds can be achieved with TIG welding, but for the best results electron-beam welding is usually chosen.
c) Are flux residues likely to cause problems? One of the reasons why gas-shielded processes are preferred for aluminium is that fluxes in oxygen fuel-gas and MMA welding corrode the metal if they are not completely removed.
d) Does the metal require special procedures to avoid the formation of cracks? Weld-metal hydrogen content in higher strength steels can be more easily controlled with gas-shielded processes than with MMA.
e) Are suitable consumables readily available? A major attraction of MMA welding is the ease with which fluxes can be formulated to meet special requirements.
f) Will the finished joint have acceptable properties? To some extent this question can be rephrased to ask if adequate control can be exercised over heat input to guarantee reproducibility. Such a requirement is often an argument for the selection of a mechanised process.
g) Do the physical properties give rise to special considerations? Copper, with its high thermal conductivity, usually needs preheating to achieve fusion. The

increased heat input in MIG welding using an argon-helium shield enables a range of thicknesses in copper to be welded without preheat.

Type of joint

a) Are butt or fillet welds required?

b) What is the position of welding? Joints in the vertical and overhead positions are rarely suitable for mechanised welding. On the other hand, many MMA electrodes are specifically designed for use in these positions.

c) Is there adequate access for the electrode, welding gun, or welding head?

d) What edge preparation is to be used? If large amounts of weld metal are needed to fill the joint, a high-deposition-rate process such as submerged arc or spray-transfer MIG/MAG welding is favoured, to keep weld times to a minimum.

e) Is an unsupported full-penetration root run required? This restricts the choice to low-current processes such as TIG and dip-transfer MIG/MAG which give good weld-pool control. Alternatively, the use of a keyholing technique such as electron-beam welding may be justified.

Production requirements

a) Are jigs, fixtures, or positioners required? These may alter access requirements.

b) Can the process be integrated with other fabricating operations?

c) What are the ventilation requirements?

d) Is there a power limitation which could preclude the use of high-current systems?

e) Can maintenance requirements be met?

15.3 Costs of the welding operation

Usually there is more than one process which meets the technical requirements of the job. This means that, in the absence of other constraints such as availability of labour, the final choice is decided by the relative costs of individual processes. At this stage we should distinguish between arc and other processes. Resistance welding is akin to the many machine-operating sequences used in engineering and can be costed as such. It is also worth remembering that resistance spot welding is more often chosen in competition with mechanical fastening rather than with arc welding. The essential components of the cost of an arc-welding operation can be summarised in a simplified formula:

$$C_w = L_w + L_a + O_h + C_c + P_m$$

where C_w = cost of welding operation
 L_w = cost for direct welding labour
 L_a = cost for associated labour
 O_h = overhead charges
 C_c = cost of consumables
 P_m = plant and maintenance cost

Direct labour

It is relatively easy to identify the direct labour element, since a welder is generally associated with one joint. There are some exceptions to this, but, even with the welding of large-diameter pipelines where two or three welders may work on the one joint, individual contributions can be readily recognised.

The biggest problem in assessing direct labour relates to the time spent actually welding, i.e. the arc time. In addition to depositing weld metal, the welder is involved in essential tasks such as setting the welding conditions, deslagging, electrode changing, and aligning the joint in mechanised welding. There are also other ways in which the welder can be occupied. For example, there may be periods of waiting for work to be delivered, or of moving from one place to another. Time must also be spent receiving instructions or discussing problems encountered during the welding operation. Electrodes and other consumables have to be drawn from the stores. Finally, there must be some relaxation time, as considerable concentration is required while the arc is being controlled in manual welding.

The welder is paid for all the time between clocking-on and the finish of the shift. In other words, all the non-arcing operations listed above form part of the total time allocated to the job and therefore are charged against the welding operation. For convenience these can be grouped as follows:

$$\text{total time} = \frac{\text{arcing}}{\text{time}} + \frac{\text{other constructive}}{\text{time}} + \frac{\text{waiting}}{\text{time}} + \frac{\text{idle}}{\text{time}}$$

To put these time elements into perspective, it is useful to think in terms of a duty cycle. This can be defined as the arcing time expressed as a percentage of the total time:

$$\text{duty cycle} = \frac{\text{arcing time}}{\text{total time}} \times 100\%$$

Since a welder is paid principally for the skill of fusing a joint and depositing weld metal, we obviously want as high a duty cycle as possible. There is a limit, however, as the welder not only needs relaxation times but must also perform the essential ancillary tasks. In addition, the layout of the welding shop often imposes limitations, with the result that the welder cannot achieve the theoretical maximum duty cycle. Similarly, the nature of the work has a marked effect on the distribution of the welder's time. Higher duty cycles can be maintained on the welding of long straight T joints compared with attaching small fillet-welded lugs scattered around a box section.

In choosing a process, we must establish which one will give the best duty cycle for the job in hand, taking into consideration the related working environment. From what has been discussed, it is clear that there is no single answer to this problem. In general, however, we can say that higher duty cycles tend to favour continuous-wire systems such as MIG/MAG and submerged arc welding. In other words, these processes show to the best advantage on long joints, especially when they are welded in situations where the waiting and idle time can be kept to a minimum. The freedom from frequent electrode changes associated with continuous-wire systems enhances the duty cycle by reducing the 'other constructive time' element. At the same time, the faster travel speeds attainable reduce floor-to-floor time and thus increase productivity. By contrast, fast travel speeds and freedom from frequent electrode changes are of little value when depositing short weld runs, especially when the joints are far apart. These circumstances tend to favour MMA welding, where the increased manoeuvrability can help to raise an inherently low duty cycle.

Associated labour

Mostly welders work by themselves, at least as far as the actual welding operation is concerned, but occasionally they need the help of another person to complete their task satisfactorily. In making an economic comparison of processes, the increased costs resulting from this additional labour must be taken into account.

Processes which reduce the extra cost to a minimum obviously have an inherent advantage. A good illustration of this is in the welding of higher strength steels, where preheat is specified with MMA welding electrodes. A change to MIG/MAG or submerged-arc welding enables the preheat to be reduced or even eliminated, since the weld metal has a significantly lower hydrogen content (see page 110). This in itself is often sufficient to justify the selection of these two processes, but additional benefits such as fewer problems in electrode storage and handling and faster travel speeds also accrue.

Overhead charges

A share of centralised expenditure must also be recovered as part of the welding costs. Functions such as both central and line management, quality control, design, stores and purchasing, sales, and general administration constitute an overhead for which production must pay. Various methods of recovering this overhead are used. One of the most common is to add to the labour costs a fixed percentage ranging from, say, 150 to 350%.

Cost of consumables

Consumable costs are probably the easiest part of the assessment which we make when comparing processes. The volume of weld metal required can be readily calculated if we know the geometry of the joint. Data is available giving the recovery rates of electrode material, i.e. making allowances for losses by way of spatter, stub ends, and so on. From this, the quantity of electrodes required and hence the cost can be estimated. Similarly, once the arcing times have been established, the cost of electricity, shielding gas, fuel gases, and cooling water can be determined.

Replaceable items of equipment may be considered as part of the consumable costs. Opinions differ on this part of the cost comparison: some companies regard contact tips, cables, nozzles, and even MIG/MAG guns as consumable items, while other allocate them to the cost of maintenance. Whichever approach is used, these replacement costs can be significant and must be taken into account in our process comparison.

Machine and maintenance costs

The last item in our formula relates to the capital cost of the welding plant and the charges made to maintain it in good condition. Mechanised plant is obviously much more expensive than a MMA welding set up. Actual prices could be quoted, but, since these change as a result of inflation, it is probably more helpful here to consider comparative costs, using MMA equipment as a base – Table 15.1.

Table 15.1 *Comparative costs of welding equipment*

Process	Comparative cost
MMA	1.0
TIG	3.5
MIG/MAG	2.5
Submerged-arc	12.0
Electro-slag	20.0

The actual cost in terms of individual welding operations depends on the amount of use made of equipment and the depreciation policy adopted by the company's accountants. Very often, it reduces to only a small hourly charge compared with the cost of labour etc.

Maintenance costs are not easy to determine in advance, and usually an estimate is made based on previous experience. In the case of MMA welding, there is very little maintenance required, especially if transformer units are used. MIG/MAG and mechanised welding systems do need regular maintenance, and an allowance must be made for this cost. The amount allocated varies from 5 to 15% of the capital cost of the plant, depending on whether items such as contact tips are regarded as consumables or as a maintenance charge.

15.4 Conclusions

It is difficult to summarise a subject as involved as process selection. Our discussions of the various aspects in the previous pages serve to emphasise the fact that there are no easy answers. In many instances our choice is a compromise, as we rarely have the freedom to adopt the ideal solution. Perhaps the most important point is that, whatever the constraints may be, a decision should not be made without first undertaking a careful analysis of all the factors involved. Nevertheless, certain key points do emerge:

a) For most of the metals used commercially in welded fabrications there is a wide choice of processes.
b) The processes currently available to the fabricator all give good-quality welds.
c) A major technical problem in choosing a process is the availability of suitable consumables.
d) Apart from welding low-carbon steel, the complexity of the best weld procedure is markedly influenced by the process, and this must be taken into consideration in our final choice.
e) Adequate skilled labour must be available. For manual processes, the craft skills are readily identifiable and are familar to those working in the industry. Gas-shielded and mechanised systems call for technician and supervisory skills which are not readily apparent and which may be overlooked in the initial comparison.

f) Cost analyses must be realistically related to the production environment and not based on a theoretical model.

Finally, it is worth noting that we are not necessarily setting out to select one process to do all the welding on a particular fabrication. Frequently, the most effective manufacture is achieved by using a mix of processes, even within a particular joint.

Figure 15.1 shows a steel cylindrical storage vessel mounted on a base.

Fig. 15.1 *Cylindrical storage vessel*

Various solutions could be found to the problem of selecting the most suitable processes. One possible answer could be as follows:

- *Joints A*
  Submerged-arc welding with a removable flux backing would offer the fastest welding and would pose few problems in terms of access.

- *Joints B*
  The root runs could be deposited by MAG welding, using a dip transfer technique to achieve uniform root penetration. The joints could then be filled by submerged-arc or flux-cored MIG/MAG welding with the cylinder rotated under the welding head. The latter process would have the advantage that joint tracking would be easier, since the welder could view the arc and correct alignment as necessary.

- *Joints C*
  MMA welding would probably be preferred for welding the nozzles into the domed ends, as access might be difficult and the weld runs are relatively short.

- *Joints D*
  Since these joints call for relatively short lengths of fillet or corner welds, it is most likely that MMA welding would be chosen.

- *Joints E*
  MIG/MAG welding with a solid or flux-cored electrode could be used here, as the joints are long enough to enable the fabricator to benefit from the faster travel speeds.

These selections have been based largely on production considerations for shop fabrication. The picture would be very different if all the work had to be done on site with no positioning equipment, in which case the answer could be to use MMA welding for all joints. The selections would, of course, have to be costed for an actual component. In the end, we must be able to say that the welding process and procedure used has given the quality required at the lowest cost.

Appendix: safety in fusion welding

In common with other operations involved in the fabrication of metals, welding creates hazards which, if not identified, constitute a threat to the safety of both the welder and other workers in the vicinity. Extensive documentation – standards, codes of practice, guidance notes and so on - has been produced with the aim of encouraging safe working practices in fabrication workshops.

Many of the hazards covered are readily recognised because they form part of general engineering or fabrication work: handling metal plates and sections or the use of chipping hammers and grinding tools. Welding also creates specific hazards which are related to the use of an arc or a fuel gas flame. Minimising the risk that these hazards pose to the welder calls for an understanding of the way in which they arise. This Appendix aims to provide the basis for the development of safe working practices related to the specific characteristics of arc and gas welding processes.

AI Safety in arc welding

Arc welding poses special problems (fig. AI) which originate from particular characteristics of the arc.

Fig. AI *Hazards in arc welding*

These hazards can be grouped under three main headings :
a) electrical
b) radiation
   i) visible light
   ii) ultra-violet
   iii) infra-red
c) generation of fumes and gases.

A2 Electrical hazards

Electric shock occurs when a current flows through the human body and causes a muscular reaction. The level of current at which this happens varies from one person to another, but, in general, currents larger than 1 mA can be felt as an electric shock. At levels above 25 mA, the heart may cease to function.

With currents between 1 and 25 mA, two effects can be produced:

a) muscle spasm, with limbs 'twitching' or kicking, often leading to secondary injury by banging against sharp objects;
b) 'hold-on', a condition in which the muscles tighten the hand grip, making it impossible to release the live conductor.

Whenever electrical equipment is used, the operator must not come into contact with a voltage which is high enough to cause a current in excess of 1 mA to flow through the body.

It is difficult to avoid contact in welding, since the electrode and plate are both exposed and can be touched by the welder. Fortunately, most arcs in welding systems operate at between 10 and 40 V and are substantially safe in themselves. Dangers arise only when higher voltages are supplied to the work; we must, therefore, determine under what circumstances these can occur and devise methods of protecting the welder.

Open-circuit voltage (ocv)

The first and most obvious source of a higher voltage is a drooping-characteristic power-supply unit which is switched on but is not delivering current to the welding circuit. The units which are used for MMA, TIG, automatic-arc, and submerged-arc welding need an open circuit voltage (see section 4.4) significantly higher than the arc voltage, to assist arc striking and achieve the correct output characteristic. Some older transformer units were provided with ocv's of the order of 100 to 110 V. While these are much liked by welders because of the ease with which the arc can be struck, they are potentially dangerous since under adverse conditions severe shock can be experienced at 100 V a.c. In recent years, positive efforts have been made to design units having lower ocvs. In the case of d.c. power supplies, ocvs of the order of 60 V maximum can be used. When operating on a.c., however, some constraint is applied by the fact that often at least 80 V is required to ensure re-ignition of the arc each time the current goes through zero when the electrode changes from positive to negative and vice versa.

A useful approach to this problem is to protect the welder against exposure to the ocv by the use of a low-voltage safety device (LVSD). Various equipments are available, but in the main they consist of a controlled contactor inserted into the welding lead. The contactor is in the open position when welding current is not flowing. The output terminal of the power supply is thus disconnected from the electrode, and only a low voltage exists between the holder and the work. As the electrode is touched to the work, a small current flows in the control unit which actuates the contactor, closing the contacts. The arc can now be struck, as ocv is available at the end of the electrode. As long as the current flows, the contactor remains closed, but when the arc is extinguished the control circuit de-energises the contactor after a few seconds' delay. The welder is once again isolated from the power supply so that the electrode can be changed without risk of shock.

With MIG/MAG equipment, electric-shock hazards are appreciably reduced, since ocv's are usually less than 50 V.

Accidental connection of mains voltage to workpieces

The welder can also be exposed to high voltages if a breakdown in insulation occurs either in the power-supply unit (not very common), in ancillary equipment, or in connecting cables with the result that the mains voltage is supplied direct to the workpiece or into the welding circuit.

Welding power-supply units are designed so that the primary and secondary circuits are completely isolated and the metal casings are connected to the mains earth. A provision for earthing the secondary circuit must be made externally to the set, and a correctly wired welding circuit should contain three leads (fig. A2):

a) *a welding lead*, connecting one output terminal on the power-supply unit to the electrode holder;
b) *a welding return*, connecting the other output terminal to the work;
c) *an earth lead*, connecting the work to an earth point.

All three leads must be capable of carrying the highest welding current which will be used. It is also important that the earth lead should be connected to an

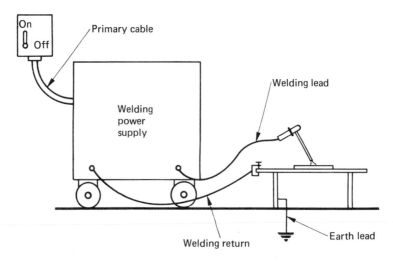

Fig. A2 *Circuit for manual metal arc welding, showing earth connection*

230

independent earth point which is capable of taking the welding current – normally the mains earth is not rated high enough for this purpose.

Check points for electrical hazards

Electrical safety requires systematic checking of details. Some of the important points to be checked are

a) that isolating switches are to hand so that the mains supply can be disconnected as quickly as possible;
b) that equipment is correctly rated and installed with adequate earthing;
c) that insulation on all cables is satisfactory;
d) that all connections are made with correctly rated components and are securely fastened;
e) that the welding lead, welding return, and earth lead are of adequate size for the currents being used;
f) that the earth lead is securely fastened to a satisfactory earth point.

Finally, every effort should be made to avoid welding in damp or wet conditions, since the danger from electric shock is much increased.

A3 Radiation from the arc

The first impression we get when looking at an arc is that it is a very bright light. This is because there is a high intensity of radiation which is within the visible range of the spectrum, i.e. of wavelengths between 0.4 and 0.75, μm. The intensity of light emitted depends on the current level and the presence of a flux.

The arc must be viewed only through a coloured filter (fitted into the welder's helmet) which reduces the light to an acceptable level. Filters are graded according to their density, and BS679, 'Filters for use during welding and similar industrial operations', recommends the best grade for a particular application (Table A1). Note: BS 679 is due to be replaced by a European EN standard.

Table Al BS 679 recommended filters

	Process and current range (amperes)		
Filter	MMA	TIG	MIG/MAG
8/EW	up to 100	up to 15	—
9/EW	up to 100	15–75	—
10/EW	100–300	75–100	up to 200
11/EW	100–300	100–200	up to 200
12/EW	over 300	200–250	over 200
13/EW	over 300	250–300	over 200
14/EW	over 300	250–300	over 200

Note Where two or more shade numbers are recommended for a particular process and current range, the higher shade numbers should be used for welding in dark surroundings and the lower shade numbers for welding in bright daylight out of doors.

231

The arc also emits ultra-violet and infra-red radiation with wavelengths less than 0.4 µm and greater than 0.7 µm respectively. These are outside the visible range.

Ultra-violet radiation is damaging to both eyes and skin. Exposure of the tissue of the eyeball to ultra-violet radiation will produce a condition known as 'arc-eye'. This is characterised by a painful gritty feeling under the eyelids, watering of the eyes, and an inability to tolerate light. Even a brief flash of ultra-violet radiation from a high-current welding arc is sufficient to cause severe arc-eye. The filters used must absorb as much as possible of the radiation with wavelengths of less than 0.4 µm. At the same time, attention must be paid to the effect of ultra-violet radiation on the skin. Small amounts of radiation simply result in a suntan effect, but prolonged exposure causes severe burning, and the welder should wear suitable protective clothing to cover any areas of skin which might be exposed to the arc.

While specific forms of cataract of the eye have been attributed to the effect of regular exposure to infra-red radiation, it appears that the risk is relatively small at the levels involved in welding. Nevertheless, as an extra precaution the filters specified by BS679 will absorb infra-red radiation, thus ensuring that there is little danger of the welder's eyes being affected.

Ironically, the workers who probably run the greatest risk of arc-eye are not the welders but those who work in the vicinity of the welding operation. Although the intensity of radiation falls as we move from the arc, it can still be high enough to cause damage at some distance away. A further hazard is created by the reflection of the radiation from polished metal surfaces, white coats, and so on. Wherever possible, the welding operation should be surrounded by screens, and ancillary workers and other personnel should be warned not to view the arc with a naked eye.

Check points for radiation hazards

To protect personnel against hazards arising from radiation from the arc, check that

a) the welder is using the correct grade of filter and a heat-absorbing filter is fitted where necessary
b) the correct protective clothing is being worn;
c) the welding operation is screened as far as possible to prevent radiation from reaching other work groups;
d) ancillary personnel are warned of the hazards arising from viewing the arc with a naked eye
e) first-aid treatment is immediately available for anyone who has been sub-jected to a 'flash' from the arc.

A4 Production of fume

At the temperatures existing in the arc, fume and dust particles are readily produced. These can be carried into the zone around the welder's face by convection currents of hot air rising from the arc. The extent to which these particles constitute a health hazard depends on their chemical composition. The

232

maximum concentration to which a welder can be exposed must be specified for each chemical compound produced as fume by the arc. This concentration is known as the occupational exposure limit (OEL), and each year the Health and Safety Executive in the UK publishes a list of OELs for substances commonly found in industrial atmospheres. These are not statutory limits but provide the basis for an assessment of the acceptability of an atmosphere.

The actual concentrations of pollutants that exist in the atmosphere in the area around a welding operation are measured by drawing a known volume of air through a filter. The increase in mass of the filter is determined, and the total amount of fume is calculated in mg/m^3. The fume is then analysed chemically to find the relative amounts of individual compounds, and the results are compared with the relevant OEL's.

Measurements are usually made in two areas (fig. A3). The first, known as the breathing zone, represents the atmosphere which the welder inhales and since it is closest to the arc, it normally has the highest fume concentration. The other is the region surrounding the welding operation and gives the *background* level, which is of importance in considering whether it is safe for others to work in the area.

Measurements made in these two positions are influenced by four principal factors: the type of process, the current being used, the presence of surface

Fig. A3 *Breathing zone and background level in arc welding*

coatings, and the type of site in which the welding is located. The last mentioned is most important, since to a large extent it determines whether the background level builds up. In open-air operations there is normally natural air movement which helps to disperse the fume. On the other hand, in a factory, although there may be some air movement in the welding shop, fumes can collect at ceiling height, giving a blanketing effect, while hot-air blowers used for heating may simply recycle the fume. The worst case is welding in a confined space, such as inside a tank, where the fume level can rapidly exceed the recommended limit.

Three approaches can be taken to the reduction and control of fume levels. Firstly, the atmosphere inside the welder's helmet can be removed and replenished by fresh air. Special helmets can be purchased for this purpose, and these probably provide the best method of protecting the welder in normal workshop situations, but it must be remembered that other people in the shop may still be exposed to a polluted atmosphere.

Secondly, the fume can be removed by placing an extractor duct in the vicinity of the arc (fig. A4). In theory, this presents the full answer to the problem since, if the fume is eliminated at source, the atmosphere in the work area will be free of pollutants. In practice it is difficult to achieve the full benefits, since the duct must be moved frequently during the operation, especially as it works at its maximum efficiency only over the central area.

Nevertheless, local extraction sensibly used can make a significant contribution to the reduction in the level of fume in the atmosphere surrounding the welding operation.

Finally, overall ventilation of the shop by way of, for example, roof extractors can help to protect workers not directly connected with welding, such as fitters, crane drivers, and so on.

Fig. A4 *Local extraction of welding fumes*

A problem frequently encountered in planning ventilation for welding shops is that of heat losses. As the number of air changes per hour required to reduce the concentration of fume to the OEL is usually large, systems which simply vent to outside atmosphere will draw large quantities of unheated air into the shop. While this may be acceptable in the summer months, at other periods in the year it can cause a dramatic increase in fuel bills. In these circumstances, recirculatory systems with filters to remove the dust and fume are worth considering, since the increased expenditure on installation may well be compensated for by savings in heating costs.

Check points for fume hazards

It is important to remember that not all fume can be seen, and precautions should be taken even when there is no visible warning of the hazard. Check that

a) material being welded does not produce compounds which have a low OEL;
b) surface coatings will not decompose to give poisonous gases or fume;
c) fluxes do not contain compounds with low OELs;
d) local extraction can be used effectively and is not hindered by the shape of the work;
e) local-extraction systems do not exhaust into the shop in such a way that they produce a hazardous concentration at some point remote from the weld area – units incorporating filters are to be preferred;
f) background levels of fume, especially underneath the roofing, should be kept low by the use of general ventilation.

If there is any doubt, the atmosphere should be sampled and analysed.

A5 Gaseous contaminants

Gases are produced in the arc typically by decomposition of fluxes or by the effect of ultra-violet and infra-red radiation on air.

Ultra-violet light from the arc can cause oxygen in air to rearrange itself molecularly into ozone:

$$3O_2 \longrightarrow 2O_3$$

Ozone is very active chemically and, if inhaled, produces severe irritation of the lungs. This results in a loss of lung capacity and an inability to make even moderate physical exertions. In contact with solids, ozone reverts to oxygen, which means that it can be readily removed from air by passage through a filter. In this respect, ozone is unique among the gases found in welding atmospheres, since the others cannot be filtered out. Oxygen and nitrogen also react under the effects of heat and ultra-violet radiation to form nitrogen oxides which irritate the lung and, in high concentrations, can induce cyanosis (blue discolouration due to lack of oxygen in the blood) followed by death. A third toxic gas found in welding atmospheres is carbon monoxide, which is poisonous if inhaled in sufficiently large quantities.

Other gases met with in arc welding usually have a low toxicity, but it must be remembered that in TIG and MIG/MAG welding the gas supplied to the arc area will displace air. In an enclosed space there is a risk of asphyxiation as the air is diluted or replaced by the shielding gas. Air must be replenished by fresh supplies piped into the enclosed area at the same time as the diluted atmosphere is extracted (fig. A5).

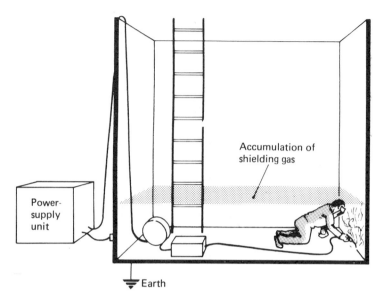

Fig. A5 *Build-up of accumulation of shielding gas at the bottom of an enclosed space*

Check points for gaseous hazards

Precautions taken to reduce the fume hazards in areas surrounding operations also remove gases produced by the arc. Additional checks are concerned with problems arising from concentrations of the shielding gases which, being heavier than air, may settle at floor level. Check that:

a) extractors are located at or near floor level if there is a risk of build-up of shielding gases;
b) effluent from extractors is vented to open atmosphere away from the workplace; nitrogen oxides are not removed by filtering and pollute the atmosphere around the exhaust pipe;
c) accumulated gases at the bottom of tanks and similar enclosed spaces are removed and replaced by fresh air;
d) resuscitation equipment is available in the event of the welder suffering asphyxiation.

As a general rule a welder should always be supervised when working in confined spaces, and provision should be made for rapid removal in the event of injury.

Vapours from degreasing fluids

These require special mention since many are decomposed by the action of ultra-violet radiation to form phosgene or other poisonous gases. It is important that degreasing operations using solvents such as trichlorethylene are not carried out in the vicinity of welding operations, in case the vapour is drawn into the arc area. Similarly, work which has been degreased in this way should be completely free of solvent before welding operations begin. Machined components containing small cavities or crevices and castings with blow holes are particularly hazardous in this respect.

A6 Safety in oxy-acetylene welding

Although the working principles of oxy-acetylene welding are similar to those of other fusion systems, it involves different considerations in terms of safety. These mainly arise from the different nature of the heat source.

Since heat is being generated by a chemical reaction, it is reasonable to assume that some of the products may present hazards. Firstly, oxygen from the surrounding air is used up in the burning of gases in the outer cone of the flame, and carbon dioxide is supplied to the weld area as a result of the combustion of acetylene. In an enclosed space, the air is diluted rapidly and the same precautions as those described for arc welding must be taken. Oxides of nitrogen are also produced and must be removed before their concentration can build up to the OEL.

Table A2 BS 679 recommended filters for oxy-acetylene welding

Metal being welded	Recommended filter for	
	Welding without flux	Welding with flux
Aluminium and alloys	3/GW	3/GWF
Bronze welding	4/GW	4/GWF
Copper and alloys, nickel and alloys, thin steel sheet	5/GW	5/GWF
Steel plate and thick sheet, pipe, etc.	6/GW	6/GWF

The light output from an oxy-acetylene flame is much lower than that from an arc, and the levels of ultra-violet and infra-red radiation are relatively small. There is no need for a face shield, and it is sufficient for the welder to wear a pair of goggles fitted with a filter of the correct density.

Although we do not have to worry about the electrical hazards associated with arc welding, the heat source in oxy-acetylene welding does pose its own additional problems in that acetylene is an unstable gas and at critical concentrations can cause explosions. Great care must be taken to ensure that there are no leaks in fuel-gas and oxygen hoses and connections. In addition, oxygen must not feed back along the acetylene hose and vice versa, since this could lead to burning in the hoses which could spread to the cylinders. This feature, known as flashback, can be prevented by the use of check valves fitted to the torch. A shut-off valve or flashback arrestor can also be fitted to the output from each regulator. The use of the correct shutting-down procedure also minimises the risk of a flashback.

237

Moving
for a world

Moving forward means developing tires with better road grip and shorter braking distances, whatever the state of the road.

CORRECT TIRE PRESSURE

RIGHT PRESSURE

- Safety
- Longevity
- Optimum fuel consumption

-0,5 bar

- Durability reduced by 20% (- 8,000 km)

-1 bar

- Risk of blowouts
- Increased fuel consumption
- Longer braking distances on wet surfaces

forward together
where mobility is safer

It also involves helping motorists take care of their safety and their tires. To do so, Michelin organises "Fill Up With Air" campaigns all over the world to remind us that correct tire pressure is vital.

WEAR

DETECTING TIRE WEAR

The legal minimum depth of tire tread is 1.6mm.
Tire manufacturers equip their tires with tread wear indicators, which are small blocks of rubber moulded into the base of the main grooves at a depth of 1.6mm.

Tires are the only point of contact between the vehicle and road.

The photo below shows the actual contact zone.

If the tread depth is less than 1.6mm, tires are considered to be worn and dangerous on wet surfaces.

NEW TIRE

WORN TIRE
(1,6 mm tread)

Moving forward
means sustainable mobility

INNOVATION AND THE ENVIRONMENT

By 2050, Michelin aims to cut the quantity of raw materials used in its tire manufacturing process by half and to have developed renewable energy in its facilities. The design of MICHELIN tires has already saved billions of litres of fuel and, by extension, billions of tons of CO2.

Similarly, Michelin prints its maps and guides on paper produced from sustainably managed forests and is diversifying its publishing media by offering digital solutions to make traveling easier, more fuel efficient and more enjoyable!

The group's whole-hearted commitment to eco-design on a daily basis is demonstrated by ISO 14001 certification.

Like you, Michelin is committed to preserving our planet.

Chat with Bibendum

Go to
www.michelin.com/corporate/en
Find out more about
Michelin's history and the
latest news.

QUIZ

Michelin develops tires for all types of vehicles.
See if you can match the right tire with the right vehicle...

NOTES

Michelin Apa Publications Ltd

A joint venture between Michelin and Langenscheidt

58 Borough High Street, London SE1 1XF, United Kingdom

No part of this publication may be reproduced in any form
without the prior permission of the publisher.

© 2012 Michelin Apa Publications Ltd
ISBN 978-1-907099-68-7
Printed: December 2011
Printed and bound in Germany

Principal Sights

REGIONS

1	Andalucía
2	Aragón
3	Asturias and Cantabria
4	Castilla-León
5	Castilla-La Mancha
6	Catalunya and Andorra
7	Extremadura
8	Galicia
9	Madrid
10	Mediterranean Coast
11	Navarra, País Vasco, La Rioja
12	The Islands

Altas
Ferrol
Viveiro
Ribadeo
COSTA VERDE
Cabo Vidio
Cabo de Peñ
A CORUÑA
Mondoñedo
Luarca
Gij
AC 552
Betanzos
Castropol
Cudillero
Naranco
AG 55
SANTIAGO DE COMPOSTELA
Lugo
3 OVIEDO
Cabo Fisterra/ Cabo Finisterre
Sobrado
A 6
Grandas de Salime
Cangas de Narcea
Sta Cristina de Lena
Valded
Muros
La Curota
Oca
8
Pto de Pajares
Rías Baixas/ Rías Bajas
Pontevedra
Camino
Vaporquero
Coto Redondo
Oseira
Gargantas del Sil
Ponferrada
LEÓ
Baiona
Ourense
de
Astorga
AP 71
Santiago
A 66
Tui
Celanova
Verín
A 52
Puebla de Sanabria
A 6
Bragança
Benavente
N 122
N 630
4
A 3
San Pedro de la Nave
T
A 11
PORTO
RÍO
DOURO
Zamora
A 4
IP 4
SALAMANCA
N 5
A 62
IP 5
A 25
Ciudad Rodrigo
La Alberca
Alba de Torm
COIMBRA
Béjar
Candelari
Zêzere
Coria
Sierra d
N 630
Plasencia
Oroр
RÍO
TAJO
Alcántara
CÁCERES
Trujillo
7
Guadalupe
PORTUGAL
A 1
TEJO
A 5
LISBOA
Badajoz
A 6
A 5
Mérida
A 2
N 432
A 66
Olivenza
Zafra
Jerez de los Caballeros
Llerena
N 432
Sierra de Aracena
N 630
CÓRDOBA
Aracena
Medina Azahara
las Maravillas
Almodóvar del Río
Minas de Riotinto
RÍO GUADALQUIVIR
OCEÁNO
Huelva
Moguer
Italica
Écija
1
Ayamonte
A 49
SEVILLA
A 4
Carmona
Faro
Marismas del Odiel
Palos de la Frontera
Osuna
La Rábida
PARQUE NACIONAL DE DOÑANA
A 92
Desfilad de los Gaitan
ATLÁNTICO
Costa
Sanlúcar de Barrameda
Arcos de la Frontera
Olvera
Setenil
Grazalema
Ronda
Jerez de la Frontera
Ubrique
Mija
El Puerto de Santa María
Pueblos Blancos
de
Casares
CÁDIZ
N 340
Estepona
A 7
Marb
la
Vejer de la Frontera
Algeciras
Puerto Ban
Gibraltar
Tarifa
Estrecho de Gibraltar
Ceuta
Luz
TANGER

Check points for oxy-acetylene welding

Check that:

a) adequate ventilation exists;
b) the welder is using the correct type of filter glass;
c) all connections in the gas-supply circuit are in good condition and free of leaks;
d) check valves and flashback arrestors are correctly fitted;
e) oxygen and acetylene hoses are in good condition and of the correct colour:
   oxygen – blue,
   acetylene – red;
f) regulators and cylinders are connected correctly, with acetylene cylinders in an upright position;
g) all connections carrying oxygen are free of grease, since an explosion can occur if grease is in contact with pure oxygen.

Storage of gases

Gases commonly used during welding are not toxic in themselves, and the usual problem is asphyxiation where concentrations of these gases are allowed to dilute the atmosphere to such an extent that there is insufficient oxygen for breathing. On the other hand, hazards are associated with the storage and handling of gases. Gases can be stored either in cylinders which are transported around the shop to be near to the appropriate welding site or in bulk containers from where they are distributed by means of pipes. The latter system creates fewer hazards and can make a positive contribution to improved safety.

Flammable gases such as hydrogen and acetylene are stored in red- or maroon-painted cylinders and have left-handed threads on connectors. Where hexagon nuts are used, the corners are notched to assist rapid identification (fig. A6).

Cylinders for non-flammable gases, such as oxygen and nitrogen, are painted black, while other gases such as carbon dioxide and argon may be identified by grey or blue paint. All cylinders used for non-flammable gases have right-hand threaded connectors fitted with unnotched hexagonal nuts.

Chamfered or grooved

Plain

Left-hand thread (combustible gas)

Right-hand thread (non-combustible gas)

Fig. A6 *Notched hexagon nut used for fuel gas connections compared with plain nut for non-combustible gases*

Gas cylinders should always be stored carefully. Apart from acetylene and carbon dioxide, the gases inside are at pressures of the order of 170 bars, and care must be taken to prevent damage to the valves. A cylinder with a broken valve can behave like a jet-propelled missile. The pressure inside the cylinders must not be allowed to exceed the stated maximum. This means that the cylinders must be stored in a cool place and must not be used close to an arc or a gas flame.

A number of recommendations regarding the storage and handling of industrial gases has been made over the years, and advice is readily available from the supplier of the gas.

Index